This book is for you if:

- You are interested in genius.

- You want to learn from great minds.

- You are interested in how ideas inspire more ideas.

- You are fascinated by the colossal achievements of Leonardo da Vinci.

- You want to understand Leonardo's thought processes.

- You want to find out how Leonardo's legacy is still being felt.

- You want to know how to think like Leonardo.

- You want to know how Leonardo shaped our future.

What people are saying

Since his very early days, my great friend Avio M. Petralia has studied and been inspired by the ceaseless curiosity, the trial-and-error approach, the thought process of the incomparable Leonardo da Vinci. But he has not just observed and absorbed–Avio has applied the knowledge he acquired from the Master. Now he has set down what he has learned, together with his insights alongside images of his own remarkable creativity that should delight and inspire. From all of this comes a very clear plan of action with extraordinary benefits: Go beyond where we are–observe what exists and then move forward to uncover what ought to exist.

Marshall Bird
International Lawyer, Auckland, New Zealand

This book is not merely a study of Leonardo; it is an invitation to awaken the latent potential within us, inspired by a soul that deeply revered and explored the full breadth of his humanity and creativity. It urges us to confront the vast, uncharted opportunities that lie ahead with courage and openness.

We are on the cusp of a transformation era in technology, one that challenges long-held paradigms. Avio offers a glimpse into how Leonardo navigated the boundless possibilities of his own time—the Renaissance—a period marked by unprecedented innovation and discovery. His story encourages us to welcome change not with trepidation but as a fertile ground for creativity, enabling us to release outdated ideas and embrace the new with a sense of wonder and playfulness.

When Avio graciously asked me to read this book, I found myself captivated, finishing it in a single sitting. It left me greatly inspired, reaffirming the belief that together, we can shape a more beautiful and imaginative future. Sometimes, we need gentle reminders of who we truly are. For this, I extend my heartfelt gratitude to both Avio and, of course, the incomparable Leonardo.

Bruce Dalzell Atherton
Portraitist, Painter and Digital Artist, London, UK

In Beyond Leonardo, Avio takes us on a brilliant and enlightening exploration of the very heart and soul of Leonardo da Vinci. It is on this in-depth journey that we are challenged to reflect on our own vision of and relationship to reality; while invited to unfold, discover, and potentially apply going forward the insights, creativity, and curiosity of this world-acclaimed treasure.

Ernesto Siravo
Founder of JetPerfect Foundation, Berkeley, CA, USA

More than five hundred years have passed, but the genius Leonardo da Vinci continues to inspire us today and in the future. Avio's book and its accompanying website show how da Vinci's greatest mind stimulated him to create feasible technological concepts that will bring far-reaching value to our lives. It's also a good reference for gaining wisdom.

Dongsu Jeo
Arts & Culture magazine CEO & Publisher, Seoul, South Korea

"Beyond Leonardo" book serves as a motivational guide, encouraging readers to adopt Leonardo's insatiable curiosity, disciplined approach, and boundless creativity. While the technological proposed concepts of the author in www.leonardoisback.com website are practical and engaging, an addition of more structured tools and applications could further enhance our ability to apply Leonardo's principles in our own lives.

Luca Pisetta

Entrepreneur, London, UK

The effect of Avio's long-standing, intense 'relationship' with Leonardo da Vinci is reflected in this book, which explores some Avio's 'too early' ideas about expanding and maximizing the use of advanced technology in daily life to provide more equitable financial benefits for everyone. In this book and on his website, he presents thought-provoking concepts about the near and distant future that may inspire you to reconsider and take action in various aspects of life.

Maureen Selvia Siwy

Journalist and Avio's wife

About the Author

In the Spanish language, my name, Avio, translates as 'provisions for a journey'. In Italian, it means 'an aircraft'. Both meanings are relevant. I hope I can take you on a journey through this book as I share with you my passion for the life and achievements of Leonardo da Vinci. I hope that these provisions provide satisfying 'food' for your questioning mind and also a means of rising above conventional thinking.

It has been very inspirational for me to have spent many years following in Leonardo's footsteps by living in Tuscany, the land of his birth, an origin he shared with many of the other 'Renaissance Masters'. I have spent many years of my life, in different parts of the world, working, learning, sharing and evolving into a more contemporary 'Beyond Leonardo' citizen, moving fast in the future but always with my roots deeply embedded in Tuscan origin.

Growing up in an artistic and musical family, I was very aware of his legacy and achievements. He appeared to be almost superhuman by the sheer volume of his work on so many different subjects. This has fascinated me from an early age.

Since his death over 500 years ago, I do not believe there has been anybody who has come anywhere near to matching his talents and achievements. The question is, 'Why?' By looking back at his life, I tried to answer that in this book studying his philosophy of life. By understanding what made him tick, we can apply these principles to the complex lives we lead today and better understand what tomorrow will bring.

> *The greatest deception men suffer is from their own opinions.*
> *Leonardo da Vinci*

 This book includes QR codes that will connect you to the website associated with the book. There, you can explore my approach to thinking like Leonardo. You will find ideas, visions, practical concepts, and solutions that I envision for the present and the future.

> *The framed quotes in this book, are original thoughts from The Maestro.*
> *Leonardo da Vinci. 1452 - 1519*

About the Website

Leonardoisback.com is a dedicated website linked to this book. It contains my practical contribution to that kind of 'contemporary Leonardo' we discuss in this book.

The website is a growing 'container' of ideas, visions, and practical concepts. Some of these concepts can be developed today, while others can be developed in the near future and possibly later. They all go beyond considerations of financial profitability and global market evolution.

The concepts aim to explore how we can use technology in different ways and what else can be developed to enhance our quality of life and efficiency.

Dedicated to:
My beloved mother, Nilia Petralia Carducci
And to
The very best presence of my life,
My wife, Maureen Selvia Siwy

BEYOND LEONARDO

His curiosity kick-started the future

Published by Filament Publishing Ltd
14, Croydon Road, Beddington, Croydon, Surrey CR0 4PA
+44 (0)20 8688 2598
www.filamentpublishing.com

Beyond Leonardo by Avio Mattiozzi Petralia
© ™PT. Duta Mitra Nawa, © 2025

ISBN 978-1-915465-86-3

The right of Avio Mattiozzi Petralia to be identified as the Author of this work has been asserted by him in accordance with the Designs and Copyrights Act 1988.

All rights reserved.

No portion of this work may be copied in any form without the prior written permission of the Publisher.

Printed in the UK, USA and Indonesia.

Table of Contents

	Forward	- 15 -
	Preface	- 19 -
	Introduction	- 23 -
1	Who was Leonardo da Vinci 550 years ago?	- 37 -
2	What does it mean to be like Leonardo today?	- 51 -
3	Rules, Values and Principles	- 65 -
4	Is Today The Day?	- 81 -
5	Personal Connection with Leonardo da Vinci	- 93 -
6	Shifting Perspective, Take Actions	- 109 -
7	Curiosity, the World of 'What if?'	- 133 -
8	The Power of Observation	- 147 -
9	Change	- 163 -
	Appendix	- 213 -
	A Letter from the Future	- 231 -

QR codes, links to the 'Spirit of Leonardo' website

a	New Global Dimension	- 46 -
b	New Creative Directions	-72-
c	A New Citizen of the World	-87-
d	The Spirit of Leonardo	-92-
e	The Value of Life	-108-
f	A New Way of Life	-132-
g	Our New Homes	-145-
h	The Art of Portrait	-157-
i	Time Machine	-162-
j	The Architecture of the Future	-200-
k	Real and Virtual all in one, and The Cube	-212-
l	Leonardoisback.com store	-228-
	The Next One Hundred Years	-237-

Forward

Where do we stand in our evolution? Can we take more control of our destiny and future? Can we continue to dream using our independent imagination? Who are we today and who do we want to be tomorrow? Ultimately, what matters is what can we gain from this evolution and how can we achieve it?

This paragraph is at the end of the book and well summarises the main purpose of it. In this enlightening work, you find not only historical and philosophical considerations, but practical suggestions as well, on how to live a modern life without forgetting that to have a fulfilled life, one cannot focus only on finding solutions to practical problems but should aim to feel truly complete as a person, having in mind Leonardo as a motivating example.

The book, you have already understood, is not intended as Leonardo da Vinci's biography. You could almost consider it a plunge into the very Leonardo himself. His presence, his greatness, permeates this work, but constitutes more a device to stimulate our own progress in life, how to proceed along its path, and how to make this path satisfactory with yourself. The need to think as Leonardo did, is there as the only way to salvage the world from a catastrophe.

It is full of hope and optimism even confronting difficult times. It is also a book about Art, written by an artist; It constitutes a glimpse, but deep, of the significance of Art and the world of Art. At the same time, it is about the 'genius' that goes with Art.

The author is an artist; nobody can deny this simple fact! But he is an artist that is completely embedded in the present and the future; he has been accustomed to the most advanced techniques and he has used (and uses) all of them, without ever forgetting his human nature, his soul and the soul of his contemporaries. Like Leonardo, he is a true Renaissance Man, a man with a Vision and capable of going into details about this vision to explain it to his contemporaries.

At the same time, he has a serene soul while also being restless. As Leonardo, he is always confident in himself and his qualities. Without boasting, he never under-evaluates himself and makes use of all the available modern techniques while underlining the importance of stopping, from time to time, to think, to physically take notes, to make good use of the silence; sometimes to look for silence to be able to reach the proper concentration and to develop, out of it, the new thought indispensable to face modern times' challenges.

Even with my background completely different from that of the author, I feel sharing almost every step of his reasoning! I deem this book to be an important addition to better understanding today's world; it can be considered a manual to life so that life itself can be complete, satisfactory and eventually, well within a frame of sustainability. It is all a choice for us all. With the right choices, mankind has not only an opportunity to escape disaster but a great future and we can be instrumental to it... just try to think like Leonardo.

Paolo Bembo
Navy Admiral
Journalist and former director of various magazines.

Preface

From the moment we wake till the moment we sleep, we are the receiving end of a constant flow of messages and information. It seems that the world is conspiring to find innovative new ways to grab our attention and sell to us. The pace of communication is also increasing in speed. The quest for our attention never lets up.

As a publisher for the past quarter of a century, I confess to being one of many trying to do the same and to grab your attention. Please forgive me! I love books and can't help sharing news about them.

However, when I came across this book, it stopped me in my tracks. I suddenly realised how little I actually knew about this amazing person, Leonardo da Vinci. I didn't expect to learn much, after all, he lived some 500 years ago, centuries before even the internet. He had never even heard of SPAM. How could he be relevant to me? Then I started reading!

Today, we use the word genius very glibly. Mainly to describe somebody who thought of something before we did. But that disrespects somebody, like Leonardo, who richly reserve that title and many more. He was a true genius but uniquely, in multiple disciplines. That is very rare.

The inspiration for this book is drawn from the amazing life and achievements of Leonardo di ser Piero da Vinci. He was a painter, sculptor, draughtsman, engineer, stage designer, architect, musician, anatomist, naturalist, physicist, astronomer, cartographer, poet and polymath. Born in Vinci, Italy in 1452. He is widely recognised as the 'father' of such disciplines as embryology, geology, architecture, and half a dozen disciplines besides. However, the one thing that you might not have realised was his influence in the field of Artificial Intelligence, in an era before computers, before the internet and even before electricity! How could this be?

In fact, the concept of Artificial Intelligence goes back thousands of years to the ancient philosophers. The word 'automaton' comes from ancient Greek, and means 'acting of one's own will.' One of the earliest records of an automaton comes from 400 BCE and refers to a mechanical pigeon created by a friend of the philosopher Plato.

Leonardo da Vinci wrote extensively about automatons, and his personal notebooks are littered with ideas for mechanical creations ranging from a hydraulic water clock to a robotic lion. Perhaps most extraordinary of all is his plan for an artificial man in the form of an armoured Germanic knight.

The design notes for the robot appear in sketchbooks that were rediscovered in the 1950s. Leonardo is said to have displayed the machine at a celebration hosted by Ludovico Sforza at the court of Milan in 1495. The robot knight could stand, sit, raise its visor and independently manoeuvre its arms, and had an anatomically correct jaw. The entire robotic system was operated by a series of pulleys and cables. The robot is described as being clad in German-Italian medieval armour and is able to make several human-like motions.
In 2002, NASA roboticist Mark Rosheim used Leonardo da Vinci's

scattered notes and sketches to see if he could create his own version of the 15th-century automaton. The Rosheim knight proved fully functional, suggesting that Leonardo da Vinci may very well have been a robotics pioneer. This is just one example of the applied imagination and skill of Leonardo. There are countless more!

Avio wrote this book to share his passion and appreciation of Leonardo's helping you to open your mind and discover creative solutions to today's problems, as viewed through the prism of his thinking.

It has been my privilege to have met and worked with the author, Avio Mattiozzi Petralia, who himself is no slouch. Avio also has an amazing brain and is also an architect, painter, artist, designer, and concept inventor. He is passionate about Leonardo, as this book will demonstrate.

My wish for you is that you too will be inspired to apply Leonardo's creativity and problem solving to those things in your life that require to be 're-framed'. I guarantee that by applying Leonardo's thought processes, you will not be disappointed.

Chris Day
Managing Director of Filament Publishing Ltd

*There are three classes of people:
those who see, those who see when they are shown,
those who do not see.
Leonardo da Vinci*

Introduction

What would it be worth to you to interview the world's most successful people and find out what the key decisions they made which contributed to their success? People like Michael Bloomberg, George Soros, Mark Zuckerberg and Elon Musk have all developed personal philosophies, based on all their experiences in life and business. Their cumulative experiences have guided the decisions they have made and have brought them to where they are today. Of course, their knowledge was important, but it was their philosophy that gave them the filter through which to assess the decisions they made.

I was born long before today's social media influencers and YouTuber idles. I have watched the world grow and evolve in far gentler times. I have seen wars take place, financial crises, pandemics and global health issues. I have watched people in the public eye who should have known better, do some incredibly unexpected foolish things. I have been an observer of life with an open mind and a positive outlook. I have always been on the lookout for role models and people with unique qualities.

This is how I came across Leonardo da Vinci at an early age. Living in Italy, it was no surprise that I would be attracted to him and his genius. He was unique in his time, and over 500 years later, his qualities and achievements have more than stood the test of time. Considering that his life predated our digital world, he did not

have any of the technical advantages we boast of today. He had no internet, no search engines, no smartphone and no apps. How could it have been that he achieved so much on his own? This book seeks to find out what we can learn from that in today's world.

I also admire Leonardo on a personal level. As a painter, musician, architect and visionary myself, I can see the level of perfection he strived for in everything he did. To me, he was a true citizen of the world and had a larger impact than any place or country he happened to be in at any particular time.

As you read this book, read between the lines to find out the hidden inspirations from the logic that Leonardo applied to every thought that he dwelt on, and every solution he created. That logic and philosophy are just as relevant today as they were then.

In fact, it is incredible that hardly a week goes by without a story appearing in the news about Leonardo da Vinci. After over 500 years since his death, you would have thought that everything that could have been said about him would have already been said. Yet recently a new story linked to his painting of the Mona Lisa appeared. The headline read 'After 500 + Years, X-Rays Have Revealed An Amazing Secret Inside The Mona Lisa'.

A minuscule fragment of paint from Leonardo da Vinci's Mona Lisa painting has been analysed and scientists have discovered new clues to Leonardo's oil-paint mixture featuring plumbonacrite. A rare compound that likely indicates the presence of lead oxide powder. The addition of this powder allowed for a thicker oil-based paint that could dry quickly. So, was Leonardo also a chemist?

This new type of paint mixture then endured as the norm for centuries. And makes the news today! I wonder how many people will be creating news headlines so long after their death? Very few!

I feel a great affinity with him, not just because we are both from Italy, but also, in our different ways, we are both citizens of the world in the widest sense.

As I embarked on the journey of discovery writing about Leonardo da Vinci, I had no idea how relevant and topical he remains to this day. He has wisdom that transcends the centuries.

Leonardo da Vinci's house is as it was five centuries ago. The essence of simplicity, the perfect place that immerses in nature, in silence and harmony. Any human evolution would probably not offer a better environment and atmosphere for becoming the Genius that Leonardo was. The perfect laboratory that connected him with his main source of natural elements and events to observe and study.

> *Two weaknesses leaning together create a strength. Therefore the half of the world leaning against the other half becomes firm.*
> *Leonardo da Vinci*

Vinci is a small town in the Province of Florence, where not many things have changed since Leonardo was born. It is a deeply medieval, typical Tuscan atmosphere where humans have developed a sophisticated yet very simple and genuine harmony with nature. Everything is made of terracotta, stone, wood, and iron, the same materials Leonardo used, even when thinking of the most futuristic machinery.

As a well-spent day brings happy sleep, so a well-spent life brings happy death.
Leonardo da Vinci

This is the Tuscany that Leonardo knew so well, in the province of Florence today, and under the dominion of the Medici family 550 years ago. The machinery, constructions, products, and everything people use have all been created through hundreds of years of evolution, of what initially, and still often today, are processes of manipulation of natural products. Leonardo da Vinci was the genius we know, just using natural resources.

> *Nature is the source of all true knowledge. She has her own logic, her own laws, she has no effect without cause nor invention without necessity.*
> Leonardo da Vinci

In Leonardo da Vinci's time, Tuscany was certainly one of the most prolific regions in Europe in terms of nursing great minds and characters. On the left side, you can see the native home of the navigator Amerigo Vespucci, who in the beginning of 1500 sailed in the South coasts of America, the continent that has its name inspired by his name. Vespucci was the first to understand that this land was not the end part of Asia, but a completely new continent. Also in the same period, there was another well known Tuscan explorer, Giovanni da Verrazzano who discovered the Bay of New York.

> *He who loves practice without theory is like the sailor who boards a ship without a rudder and compass and never knows where he may cast.*
> Leonardo da Vinci

Chapter 1

Who was Leonardo da Vinci 550 years ago?

Leonardo was born on the 15th of April, 1452 in Italy. Today, we remember him as a great painter, draughtsman, architect, engineer, scientist, inventor, theorist, sculptor and visionary. He was a genuine Renaissance man with everything that this means. In recent times, we have become a world of specialists, with each of us knowing more and more about less and less. Leonardo was different. As a polymath, he could become an expert in any discipline he chose.

He was multi-talented, multi-skilled and deeply appreciated nature in all its forms. It was his curiosity that drove him to understand the world around him, to research, discover and invent. He had time for little else.

Leonardo was a great observer and was constantly sketching the things that interested him. He believed everything was somehow connected and a part of something more significant. He tried visualising the 'full picture of creation by the Creator', where everything was related to everything else.

His ambition, throughout his entire life, was to understand the way that things worked. The huge power of nature, the universe and all the people and creatures that were a part of it. Such was the scale of his interests that, even today, many of his personal papers and drawings are still revealing hidden gems. He also had an incredible imagination and talent for expressing this on paper and canvas. The legacy of his art is unsurpassed and includes the amazing Mona Lisa, now in the Louvre Museum in Paris.

As a true visionary, he could think and visualise far beyond what most people can imagine. He could see through things and imagine how they were on the inside. His vision had no limits and was stimulated by everything he saw. Throughout his life, his mind was razor-sharp and his legacy can be seen in museums and significant collections around the world.

In his last few days, he agonised over his achievements, asking himself, 'How many things have I left uncompleted? How many more things could I have done?' A question we should also be asking of ourselves!

Leonardo was very strict with his use of time. He recognised how precious it was and was very disciplined with what he chose to spend it on. He believed in planning each task in detail and allocating a specific time to each job. He believed that we are all in complete control of what we do and how we use our time. Using this principle, Leonardo had an impressive and efficient work rate and often took on new commissions. There was possibly a financial incentive for this as well!

Leonardo was one of the most outstanding scientists that humanity has ever seen and was always searching for clues that could explain the 'Divine formula and mechanism' that regulates everything in our world. 'The essence' of why the Creator put everything together in the way He did. As part of Leonardo's search, he would spend time talking with people at the end of their lives to try and find answers based on their experiences. Then, when they passed away, he performed an autopsy on them and opened up their bodies to find out the 'how' and the 'why' of their existence. His anatomical studies showed what detailed knowledge he had developed as a result.

With all of his diligence, research and questioning, he did learn an incredible amount about the workings of the human body. His experiments unlocked much about Nature and the mechanics of Life, but the big questions about Divine purpose and the Formulas which controlled life itself eluded him.

Today, some 500 years later, we have done so much, created so many new technologies and massively increased our knowledge, but those key unanswered questions are still there, unanswered. Since Leonardo's time, life expectancy has increased. Today, people live for great ages, which would have seemed impossible in Leonardo's time.

We have a greater perception of who we are and what we can achieve. We live in a world of opportunity and have certainly extended and expanded our capacity for doing, communicating, and interacting.

Even though Leonardo lived in a small country village, that didn't hold him back. He travelled to broaden his knowledge, experience and connections to gain inspiration–and commissions! In doing so, he created his opportunities. With life today being so much richer for us in so many ways, and with incredible technology at our fingertips, can we say with 100% certainty that our lives are any better than in Leonardo's time?

Essentially, what he was trying to discover 550 years ago is much the same as what we are trying to find out today. Fundamentally, nothing has changed. The big questions remain but we have added several new ones which are equally perplexing and thought-provoking.

Despite being surrounded by so much that is trivial and meaningless, there are still people who have the spark of Leonardo within them and the quality of curiosity that drove him to seek out the unknown. Do you have the same spark within you? How open is your mind? Hopefully, this book will help you to find out.

So, What Can Leonardo Teach Us Today?

For Leonardo, the Creator was the reason for his existence and he believed that everything that he produced and created was a way of showing gratitude to the Master Creator. He had the ultimate freedom to focus his talents on every aspect of human life, from the practical to the artistic. He was never afraid to look outside the box and to try anything. Even his most outrageous thoughts that were rubbished at the time, have later been endorsed as being viable by modern science.

Whenever we are surprised by significant changes in our world, they always highlight that the status quo has been around for far too long. Change creeps up on us.

Leonardo's greatest gift to us today is to apply his way of thinking to topical issues and problems. With an entirely new industry emerging centred around Artificial Intelligence, the scope for inventiveness and creativity is huge. This is attracting a new generation of new Leonardo.

In his day, Leonardo thought the impossible. Even the idea of flying machines or submersible craft was unheard of and yet he conceived them and went on to prove how they could work. What surprises could emerge from space; from the metaverse or cyberspace? They are all wide open for Leonardo's style of thinking.

Evolution never seems to be a smooth path, but instead a series of steps. The flat step shows the avoidance of change, and the clinging to what we are comfortable with. The sudden step is where a new technology becomes irresistible, and change is forced to happen. This has happened regularly over the centuries.

Leonardo's skill was in targeting expectations. The commissions he received from Kings and Princes, for his military machines were sold on the expectations of the impact they would have on the battlefield. Looking at his sketches of these huge machines, it is easy to see the impact they had.

He had a great skill in identifying a need and then designing a machine that would deliver it and he had no shortage of patrons who would finance him to do it. He was able to share his visions and inspire others to believe in them. Then to pay him to bring them to life. He was a businessman first and foremost!

Our lesson today is to be like him, constantly on the lookout for new opportunities in the marketplace, and look for new problems that need a solution. Then take action! Most people don't, and quickly move on to something else. They shy away from anything that looks like work and which could threaten their comfort zone.

Knowledge itself is not enough. We also need inspiration, vision and application. Leonardo was a man of action not just a theorist or simply an academic. He was an entrepreneur as well. He wanted to turn his creativity into money! His energy was unstoppable! We need to think and act like Leonardo!

Arguably, the range of opportunities today is vastly greater than in Leonardo's time. There are so many more fields of science; artistic expression, materials to work with, and forms of transport. There is the full expanse of both the oceans and also of outer space. All we need is the determination to explore them. Are our lives so comfortable that we have lost the urge to explore and satisfy our curiosity? It is someone else's job?

I believe that in another 500 years, we will still recognise, celebrate, and learn from Leonardo because of his lifetime of curiosity. Many have been inspired to walk in Leonardo's footsteps. Extraordinary minds like Albert Einstein, W.A. Mozart, Stephen Hawking, Nikola Tesla and maybe even you! We all have the same seeds of genius within us but so often we fail to recognise them, let alone water and nurture them!

What difference could you be making to the world if only you released your imagination to go and play? After all, creativity is just the brain given the freedom to imagine and play!

If Leonardo were here today, what would he be focussing on? What would appeal to his great intellect? One thing you can be certain of is that it would not be where we expect! He would draw from his past knowledge and look elsewhere. Our challenge is to try and think where that could be. Where is nobody else looking? Think like Leonardo! Think the unthinkable and discover the unexpected.

> *One can have no smaller or greater mastery than mastery of oneself.*
> *Leonardo da Vinci*

Men of lofty genius sometimes accomplish the most when they work least, for their minds are occupied with their ideas and the perfection of their conceptions, to which they afterwards give form.
Leonardo da Vinci

Leonardo da Vinci can inspire anyone to go beyond.

NEW GLOBAL DIMENSION

We need to move forward with something capable of breaking through every comfort zone. Going well beyond what most of us do today, using the full capabilities of technology the way it can deliver today.

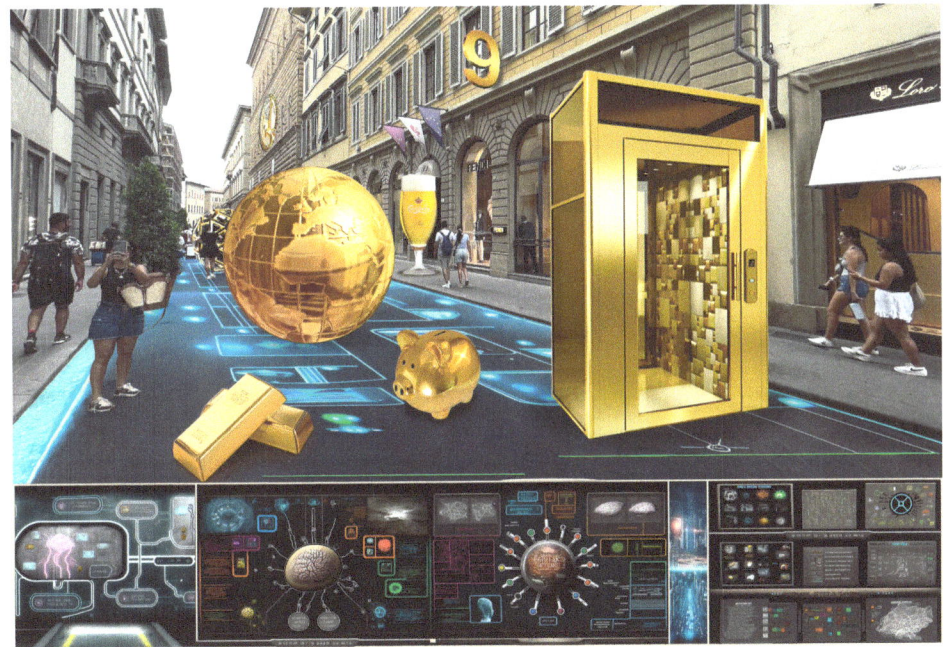

A beautiful body perishes,
but a work of art dies not.
Leonardo da Vinci

Living in 5 dimensions: height, length, width, real-virtual and time.

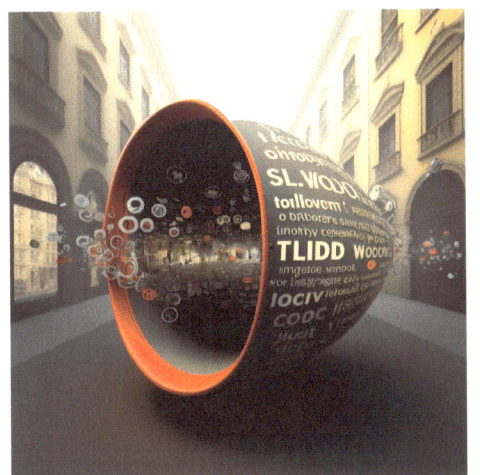
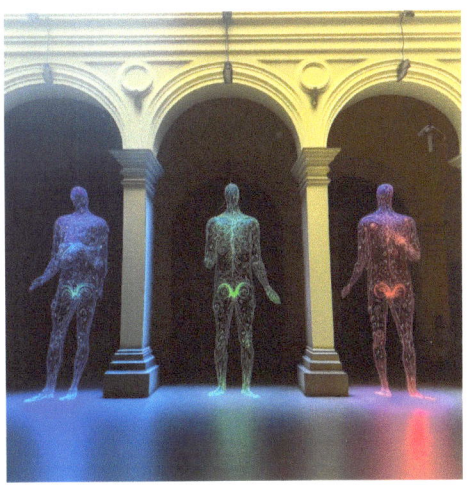

Mirroring the real world with a remote, fully connected virtual one.

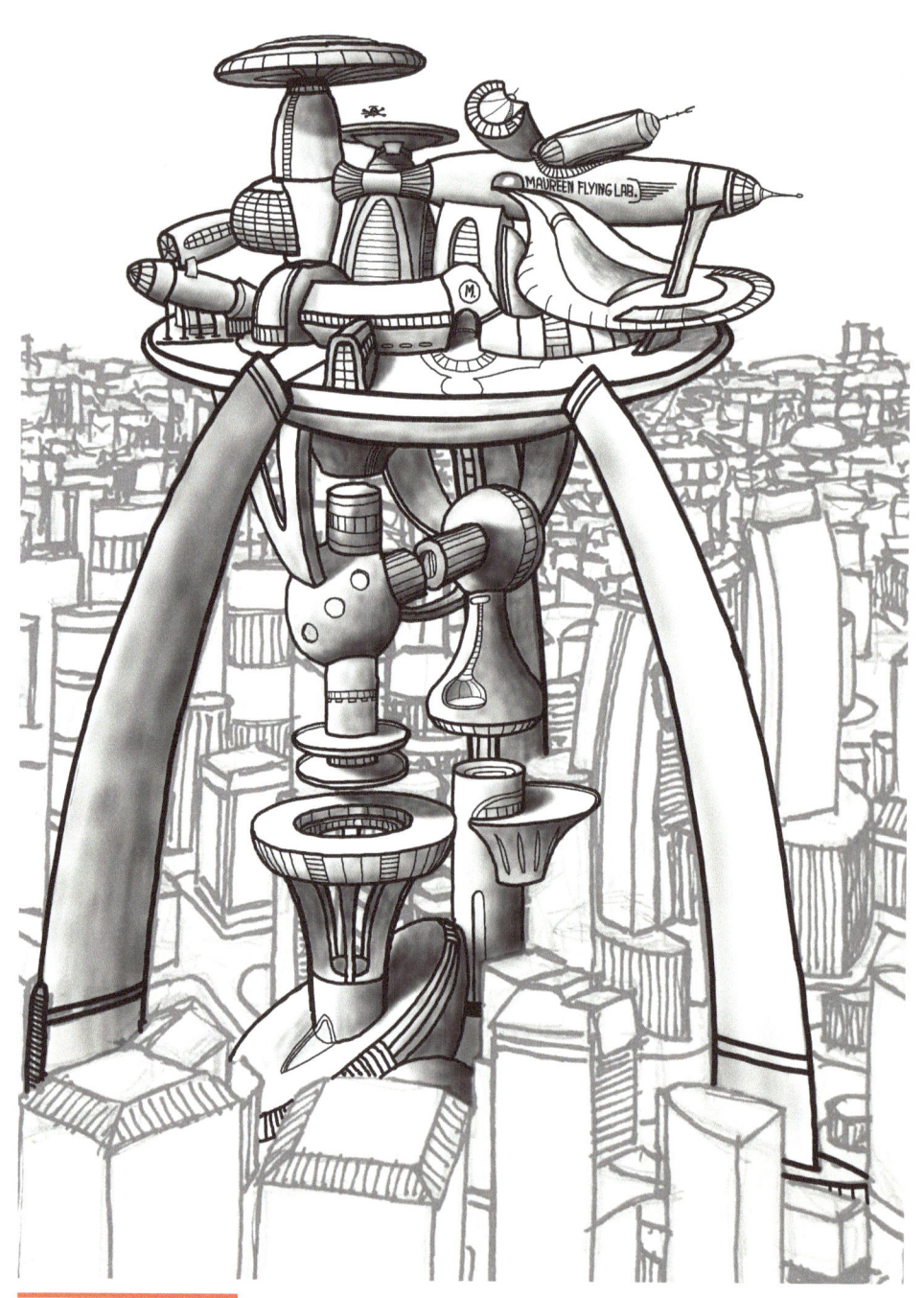
Multi-layers and levels of buildings, designs, and time.

Chapter 2

What does it mean to be like Leonardo today?

It might be difficult to relate to somebody who was born 500 years ago. We might look at what he didn't have and never knew about. The technology we have that he didn't. This would be missing the point. Leonardo's huge achievements happened without any modern tools we have grown used to, including the internet or Wikipedia! What we are talking about is his incredible curiosity and thinking ability.

Leonardo's biggest achievement was probably focusing all of his energies on his constant desire to observe, discover, and create, combined with his powerful eyesight and constantly active imagination.

He did not allow himself to be distracted by having a family of his own. This gave him the total use of his time to uncompromising focus on what he was driven to achieve.

These days, we get a few true polymaths who possess the capability to work in multiple specialisms. Therefore, he could seamlessly apply knowledge of one branch of science to another without a blink. These days, it is unlikely that a physicist will have a working knowledge of biology or chemistry. In this regard, Leonardo was at an advantage with an in-depth knowledge of all disciplines.

This was the era of specialists who focused on multiple subjects while multi-tasking. You could be an architect, painter and sculptor, engaging across various fields. Leonardo da Vinci was a prime example of someone with a large spectrum of knowledge in many disciplines, that is why he had such an advantage. He recognised and demonstrated that many things were interconnected in a much more sophisticated way!

The freedom to follow an idea, irrespective of which scientific field it touches on, is something the past can teach us. Are we working too much in compartments? The brain doesn't!

In the Renaissance, during the thirteenth and fourteenth centuries, many big inventions were still to be discovered. Art, anatomy, engineering, science and architecture were Leonardo's favourite disciplines. Perhaps his skill was seeing them as one big playground for his brain, which eventually gave him the edge.

Has everything been invented now? Is there no more need for big thinking? Artificial Intelligence would suggest otherwise. Now, more than ever, we do need to think about the impossible and dream big dreams. Nothing less! 'If people were wading through your deepest thoughts, would they get their feet wet?'

The biggest opportunity is to spot the germ of an idea before it takes root and before anyone else recognises its potential. Using tools like Google Alerts to search for stories related to selected keywords can put the latest thinking in front of you before the world wakes up. Being constantly alert for the new and unusual is also a quality of Leonardo–even though he didn't have Google!

The biggest challenge for us today is to be able to keep up with the amazing speed of evolution and information flow. In Leonardo's era, everything was in slow motion and basic. Most of the population was very limited in their studies and general knowledge. A Sunday newspaper today would contain more words than the average person in the Renaissance period would encounter in a lifetime.

In his time, Leonardo was constantly researching, writing, experimenting and creating. Most of what he was doing was misunderstood by those around him. His visions were out of scale, out of proportion, and simply not feasible, but Leonardo took no notice and was unstoppable.

Ask yourself, if you were as single-minded and driven as Leonardo, what more could you have achieved in your life so far? If you had not allowed yourself to have been distracted from your purpose and given it your total engagement, would you be somewhere different?

Even if you were to remove Leonardo's genius, his sheer energy and determination would have caused many of his achievements to happen. How many things do we know how to do but find the lure of the sofa far too strong for us to get off it?

Fatigue Makes Cowards of Us All!

There are no barriers to us being as productive as Leonardo. It is all down to the choices that we make. We all have 168 hours in every week. That is all that's available, no matter who we are! What we choose to use each hour for is completely up to us. There is a saying,

'If you want to have something done, ask a busy man!' Does he have any more time? No. But he is in better control of how he uses it.

Leonardo would be horrified at the amount of time we spend on our mobile phones and social media. According to research, we can easily fritter away an average of three or more hours daily, which is probably a huge understatement!

We are all capable of so much more. We just lack the desire and a big enough purpose. What would you be doing now if you knew you couldn't fail?

Today, the speed at which new technologies emerge is truly breathtaking, so much so that we expect a solution to every problem to be quickly found. When it is, after a short period of novelty, we quickly take it for granted as if it had always been there. It is easy to forget that it was Leonardo who, without computers, calculators, Apps or software, single-handedly invented the concept of a flying machine, parachute, and scuba gear. What could we, with all the advantages of today's technology, be creating? There is more waiting to be invented than we can imagine! Are you waiting for somebody else to take the lead in your own world? There is a Leonardo in each of us just waiting to be recognised and released. Or isn't there?

The trouble with having ideas is they require energy, time, and concentration. There will always be someone else who will recognise the problem and the opportunity. No need then to get up from your comfortable sofa! You can only create change in the world if you first make changes to yourself. If not now, when?

History will recognise the curious, the inspired, the thoughtful, the industrious and the imaginative. In Leonardo's case, we are still talking about him and learning from him half a Millennium after his death. What does that teach us?

In fact, with everything we have at our disposal, we have no excuse for not leaving our mark on the world. The biggest regret you will ever have is not for anything you may have done but for the things you did not do.

If there is a sleeping Leonardo inside you, NOW is the time to wake him and let your creativity soar! You only need to give yourself permission to open your mind and shake off all constraints and conventional thinking, as Leonardo did. It was only then that he could allow his creativity to shine through.

Common Sense

We are all born with the same five senses. They are sight, smell, hearing, taste, and touch. The organs that do these things are the eyes, nose, ears, tongue, and the skin. All the information we need comes from these sources. You would have thought that somewhere in our upbringing, we would have been taught how to use these senses to the best of our ability. Sadly not! As a result, most of us only:
Look without seeing.
Taste without appreciating.
Listen without hearing.
Smell without understanding.
Touch without feeling.

Our senses tell us only what we already know, not what has yet to be discovered. To really understand our world, we need to train our senses. We will discover so much more in the same way as Leonardo.

All our senses combine to provide our brains with an incredible amount of data. His skill was in interpreting it, understanding it, being inspired by it and adding that knowledge to his super-brain till the moment came when it was needed.

Feeding the Brain

Leonardo also had another big advantage in his life. His diet was very different to what we eat today. He was fortunate not to have a fast-food outlet at the bottom of the road in Vinci. We know today how healthy a Mediterranean Diet is. In fact, it was not until recently that we realised how healthy it is for the brain.

Professor Michael Crawford recently released a book called 'The Shrinking Brain' in which he identifies how essential the fish oils on the Mediterranean diet are important for building blocks of the brain. Without them, our brains cannot achieve their optimum size. Leonardo grew up with the perfect diet to build his brain. According to Professor Crawford, DHA and Omega-3 are the building blocks of the brain–the body does not create them, and without them in our diet, we are in danger of becoming a race of morons. Leonardo's genus was also a product of what he ate. Food and our choice of diet are just another example of 'everything is connected' because we are what we eat.

The Mediterranean diet that is found in Tuscany consists of plenty of fruits, vegetables, bread and other grains, potatoes, beans, nuts and seeds; olive oil as a primary fat source; and dairy products, eggs, fish and poultry in low to moderate amounts. This is a far cry from the over-processed meals and lack of fresh produce, we have become far too familiar with today.

What Would Leonardo Be Working on Today?

If Leonard achieved so much using the very basic tools of 500 years ago, you have to ask yourself what on earth could he achieve with the technologies we have today?

One of his passions was architecture. A century earlier, he saw how the plague devastated his homeland. He blamed poorly designed, walled cities for the fast spread of the disease. Leonardo addressed this in his proposed designs for cities, particularly, Milan. He set out to tackle the growing issues of hygiene and transportation and proposed a city built on several levels. This plan separated pedestrians and traffic and limited the height of structures to maximise sunlight and improve the quality of life. He focused on building around a centralised plan.

In addition, he was involved in redesigning the Milan Cathedral, where he proposed that a new dome be built. In his sketches, he included the plans for the dome and the layout and internal structures that exhibited his vast engineering knowledge. His plans were not chosen for the project; instead, they went to architects Amadeo and Dolcebuono. You can't win them all!

Da Vinci found much of his inspiration in nature. During his time spent in the Loire River Valley, a region known for its grand château, Da Vinci designed a double helix staircase for King Francis I, who appointed Da Vinci as first painter, architect, and engineer to the King. Whilst living on the grounds of the Château du Clos Lucé, Da Vinci drew inspiration from a snail for his model of the staircase and produced what is now thought to be one of his masterworks of design and engineering. The staircase can be found at the Château de Chambord in France.

Leonardo influenced my thinking from an early age. When I visited the Uffizi museum in Florence with my fellow art school students, I came across his work, The Annunciation. I noticed that his work was very different from all the other masters I was looking at. Comparing his work to other artists, Leonardo gave me a very different experience. His art was a collection of messages, and his techniques stood out from everyone else.

One of the greatest things I learned from Leonardo was to capture every idea I had immediately once it came to me. Inspiration is fickle and can vanish as fast as it arrives. The faintest of ink is better than the strongest of memory.

Always carry a pocket notebook and a pen. A pen is the fastest object to use. It can capture your thoughts faster than any other device. It doesn't need to be switched on, it doesn't need you to find the right programme, you don't need to open a new page. Don't get complicated! Capture each thought as it arrives and before it leaves, using a pen! When you see the thousands of sketches Leonardo made, you will understand why.

It is amazing what you can learn through close observation about Mother Nature and all the people we meet or interact with. All these nuggets of gold will be readily available to us in the future, provided we diligently captured them on paper! Leonardo certainly did.

Seeing the Future

Another aspect of Leonardo was his eyesight. In reality, it was probably no better or worse than anyone else. Still, he developed the skill of deeply studying anything he was looking at and could recall the many details most people would miss. This is particularly visible in his still-life and nature studies. With his paintings, no matter how often you look at them, you will see new details each time. A testament to his power of observation.

Sadly, my eyesight is not good, so I have had to train my imagination in order to compensate. I was the complete opposite of Leonardo in this regard. I use my imagination to fill in all of the gaps that my eyes could not see. During such a period of imagining, I started to wonder what it would be like if, somehow, Leonardo might 'return to earth' in our time. What might he think of our world today? What would amaze him? What would horrify him? What problems would he see that would prompt him to want to invent a solution? It is an intriguing thought!

Certainly, there would be a queue of wanting to interview him and get his insights into a wide range. I suspect he might see the world as having become over-complicated to the point that fundamentals become overlooked. Looking at his designs, I assume he would want to add more beauty to the world around him. Certainly, he would not be thinking small. The scale of his imagination was huge and would quickly transform wherever he was.

Leonardo would also search for the divine connections that linked everything in the universe. This was his lifelong quest. He would approach everything from a different angle. He would put the idea and objective first and then source from everything available to fulfil the objective.

Time stays long enough for anyone who will use it.
Leonardo da Vinci

Visions of different type of cities, spaces with so many humans living together.

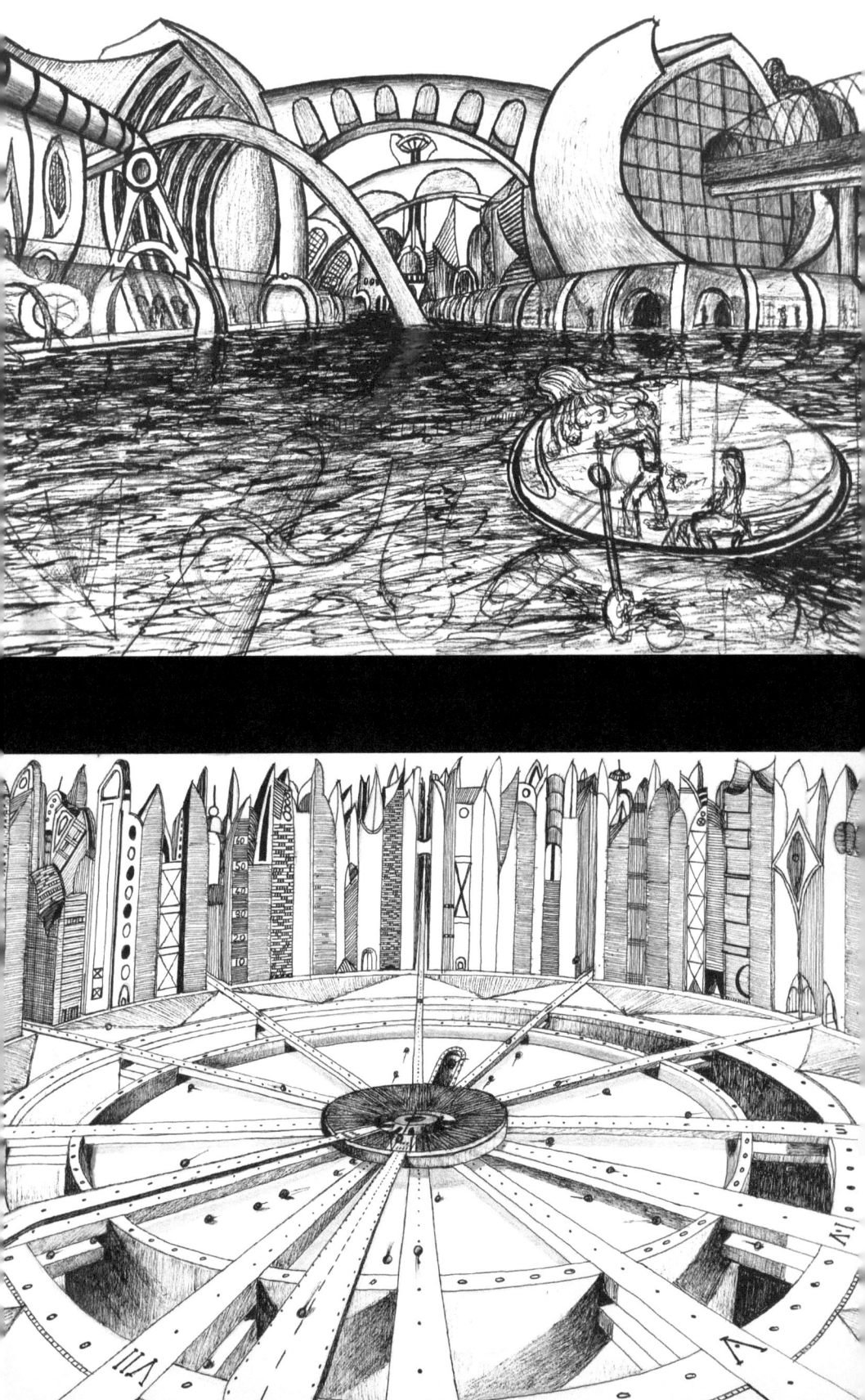

*Where there is shouting,
there is no true knowledge.*
Leonardo da Vinci

Chapter 3

Rules, Values and Principles

The speed of nature has not changed over millions of years. The seasons come and go as regularly as clockwork. The Earth rotates once every twenty-four hours to the second, with the sun rising and setting the same predictable rhythm.

These days, humans are not satisfied with this pace and would prefer everything to go much faster. Speed seems to be far more important than taking time to watch the beauty of our world unfold in the heavens and all about us.

Man's quest seems to focus on building the fastest train, the fastest plane, providing the quickest delivery service, or creating the fastest food. Never mind the quality, just give it to me faster!

I am sure that, in his day, Leonardo would have shuddered to see how much our focus has changed from quality to speed. He was such a perfectionist. For him, speed would be governed by the horse pulling the cart or how fast he could walk. And that was fast enough. That gentle pace gave him the time to appreciate nature in all its glory. Enjoying the sun blazing down on the olive groves close to his house and allowing his creative mind to develop ideas. Beauty was all around him, and everything was a part of the great Divine Order that he found fascinating.

To put things into perspective, Leonardo and the Florentine Renaissance arrived after the Dark and Middle Ages, which followed the fall of the Western Roman Empire. As the names suggested, it was not the most vibrant period in history. Society revolved around deep religious rules that governed every layer of life. Rules had to be obeyed. In any town or village, the church was built in the centre of everything. It was a visible symbol of power reinforced with the largest building in the heart of the main square. You could not ignore the Church.

In addition to the Church, there were many other sources of power. The power that was enforced with might and with the sword. This was not a time that you would want to fall out with anyone! Where the Church had religious power, Kings and Princes had the economic and secular power and were not inhibited in using it. They had the protection of living in fortified castles and had their own armies to protect them.

Then, with the arrival of the Renaissance period, Florence, guided by the powerful Medici family, woke up to art, culture and science. Fine art was a demonstration of wealth and success, and all of the nobility competed with each other to show off their paintings, statues and buildings as a sign of superior intelligence. What better time for Leonardo to be living. With all his many talents, he was perfectly placed to provide them with what they wanted! At a price! He attracted commissions from the leading families for paintings, sculptures, inventions and even weapons!

This was also the time when the foundations were laid for what has become the modern 'banking system' of today. The word bankrupt derives from that era when, if a merchant could not meet his obligations, his bench would be broken or 'bankrupt'. Banking started from strategic alliances between wealthy families in an attempt to break the power of the church and control the world of business.

The art workshop's creative and business model in Florence flourished, and orders from all over Europe were collected for building architecture projects, portraits, sculptures and paintings. Around this time, the role of the architect emerged as a specialist to cope with the surge in building projects.

Paintings were full of symbolic messages and subtle codes between people and families. Sometimes, pictures were commissioned, not just as a thing of beauty but also as a means of communication. A man might be trying to send a message to a future spouse using symbolic elements! A new industry was born of creating custom-made Art on demand. It started in Florence but spread rapidly elsewhere.

This new 'Creative Art Industry' was born and provided profitable work for many master painters, 'bottega dell'arte' (art workshops) containing painters, artisans, sculptors, wood carvers and all the other crafts that supported them. Clients could order from a menu of features and elements that could be included in a painting, all of which had a price. Not to mention its sheer size. With paintings, it was a case of 'size is everything'!

To feature a book in a painting would have different prices depending on whether it was closed and just showed the cover or whether it was open and displayed the text of contents. There was a menu that the client could choose from. The more complex a painting was, the higher its perceived value—and the cost of producing it! Art was business! Clients were prepared to pay for a high-quality result that demonstrated how cultured (and rich!) they were.

Some artists developed and offered to include perspective, the artistic use of shadows and lighting effects. This was a competitive marketplace, and you had to be wealthy to be a part of it. There was competition in quality, cost, effectiveness, size, design, and everything else. This was cutting edge in the world of art and it was booming.

Today's appreciation for art dates back to the influence of the exciting painters of this era, and Leonardo was up there with the best of them. The Florentine Renaissance inspired the evolution of art up to now. It was the turning point in making art a science of beauty and not only as a representation of religious subjects. Great artists such as Pablo Picasso, Amedeo Modigliani, Andy Warhol, and many more we can think of exist because of that significant turning point.

Leonardo was born exactly in the middle of this exciting period and at the peak of this 'intellectual and commercial revolution'. He became a role model for many other artists, and we are still admiring his work centuries later. Of course, there were many other masters like Michelangelo Buonarroti, Sandro Botticelli, Raffaello Sanzio, and great scientists like Galileo Galilei, but it was Leonardo who combined artistic and creative skills along with scientific, experimental engineering, and the big vision in his work was an unbeatable combination. His imagination offered futuristic ideas such as tanks, helicopters, planes, and submarines, none of which existed at that time, but just might if a sponsor could be found to fund them. That was the key difference between him and all the others.

As an entrepreneur and a businessman, Leonardo was well able to market himself in a powerful and original way. When talking to his sponsors and clients, he would inspire them with his vision for creating the impossible. He was a 'professional inventor', always looking at the unknown to find solutions. He often made extravagant claims to attract a sponsor, not all of which proved to be correct, and with some disastrous results! Does this maverick still have anything to teach us today?

Surprisingly, because of his extensive knowledge of so many of the sciences, and his analytical brain, he was able to see links between them that, because of specialisation, we might miss today. Making connections between diverse and unconnected concepts was one of his greatest skills.

In my personal opinion, Leonardo has much more to offer people today than you might imagine. Just compare your own qualities and thinking abilities to Leonardo's. How do you rate yourself with imagination and creativity?

Leonardo was in the business to give his sponsors and clients what they wanted. In fact, many of them didn't know what they wanted until he painted a compelling word picture to describe it. His ability to inspire was huge, even if what he could envision didn't quite exist yet. Leonardo was a great listener. Listening is an art. So many of us use the time when the other person is talking to come up with what we want to say next. Leonardo had the skill of deep listening and of committing that conversation to memory to revisit later. A great skill.

Leonardo wasn't afraid to fail. He would put 100% of himself into every project and trust his skill and experience to make a success of it. He didn't always succeed. The lesson to take from this is, if you knew you couldn't fail, what would you be doing differently now? So often it is a lack of belief in ourselves that causes us to be timid when we should be strong. Leonardo believed 100% in his abilities and ideas.

Leonardo had the confidence to build relationships with people at the highest levels of society and to negotiate, knowing the value of what he brought to each project. He was unique and charged accordingly. He never undersold himself! Do you know your own value?

Leonardo also did something very simple, yet very valuable. In his notebooks, he wrote down everything. Every thought, every idea, everything he discovered, everything that passed through his mind. Even today, people are still discovering new things in the vast quantity of notes that he created and left behind.

Leonardo had a great interest in military strategy and the design of weapons. He had skills as a mechanical engineer that were way ahead of his time. No wonder the warring princes were particularly interested in him and what he could invent to give them an edge. He was also an amazing dreamer capable of visualising and committing many of his visions to paper. He was an aviator, as yet, without a flying machine, but he really deserved to have one! In his days, he was a big traveller and a specialist in subterfuges and the ability to create a mystique surrounding him and what he was working on.

Now, what about you? List down your personal principles, values, qualities, skills and ambitions. Are you achieving your full potential? Which of your qualities have you yet to unleash? What could you be doing if you really wanted to? Let Leonardo's many qualities inspire you to even greater things. Life is not a rehearsal!

NEW CREATIVE DIRECTIONS

At the beginning of the last Century, the creative industry was flourishing, with many movements, innovative visions and ideas. Thanks to the rapid evolution of technology and globalisation we have today, we can explore and develop new forms of artistic expression.

How it is more difficult to understand nature compared to a book by a poet.
Leonardo da Vinci

On the right page: Four or more seasons together.

A dream of Florence.

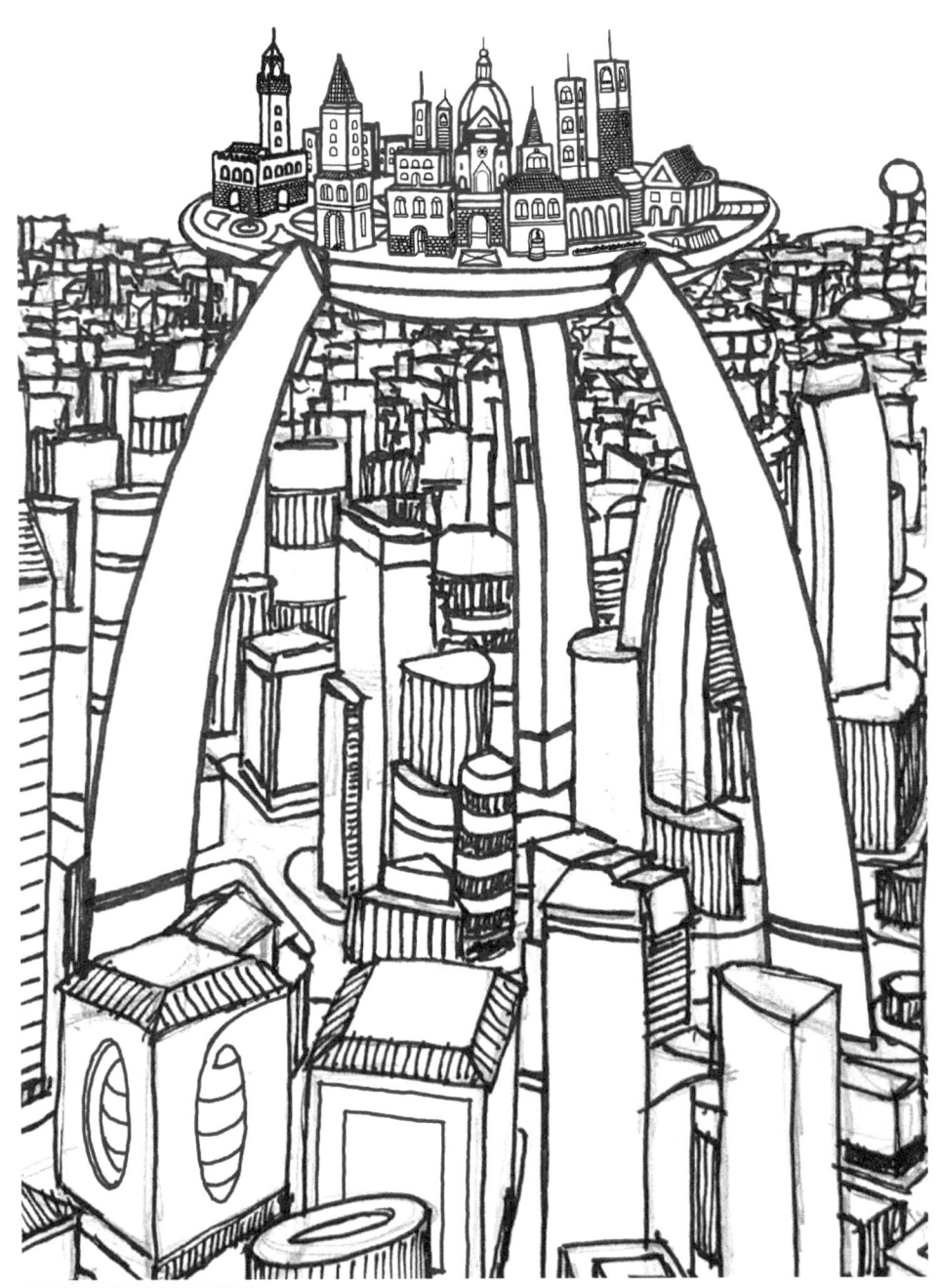
Florence-flying above a futuristic city.

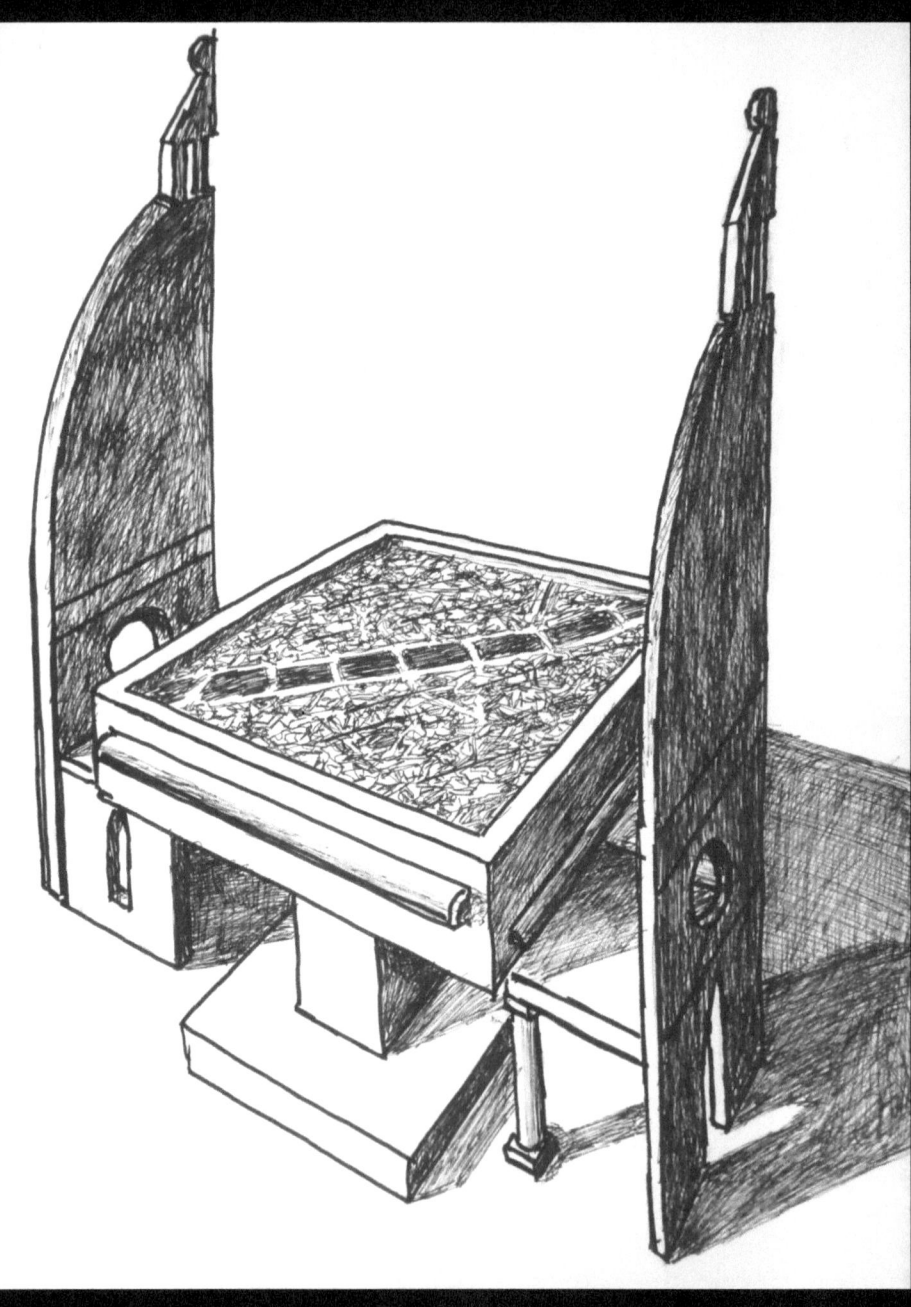

A multi-functional everyday station.

A table, an RGB light, a personal memory and a mood status board.

A dream of Paris with a vibrant sky, colours, and life in constant motion.

Chapter 4

Is Today The Day?

Let's look at the big picture. We are now in the 21st Century, so let's say that the world as we know it is 21 years old. So, in human terms, it has just moved from being a teenager to an adult. It has yet to reach maturity or to reach maturity in later years. So, where do you believe that humanity will be at thirty-five?

What will have been learnt, discovered or invented in those centuries? Will our planet still be habitable? Will global warming have engulfed us? If we are looking to change the world at 21, will there be a world left to save? What will have been our contribution towards it? Right now, we need all our best brains focused on this. We all need to think like Leonardo!

Leonardo said: If you are alone, you own 100% of yourself and your time. That was true 500 years ago and is still true today. He allowed few distractions into his life and, as a result, was very productive. Our biggest limitation is our lifespan. It is too short! There is much we want to do and so little time to achieve it! Looking at Leonardo's massive contribution, who could come anywhere close to it? And yet, in his last hours, he was criticising himself for not achieving enough. I resonate with his wish to be a 'contributor to evolution' in every walk of life.

As you might have already gathered, Leonardo has been a big inspiration for me. I have studied his many qualities and am truly impressed by his work's beauty, complexity, creativity, and vision. He was a practical visionary solving the problems of his day but was also capable of thinking far beyond his time.

Today, we take for granted our ability to share thoughts, ideas, and communication worldwide in seconds. We can get replies from people almost immediately, which enables us to work collaboratively with them wherever they may be.

I am certain that Leonardo would be particularly jealous of this. What more could he have achieved if he had been able to communicate with people so quickly and to crowdsource from other great minds? What might he have achieved if he were able to converse with other geniuses regularly? Such conversations would be eye-opening!

Today, we have many communication channels, but do we use them productively and intelligently? Or do we use them for gossip and trivia? If Leonardo had been able to talk to anyone in the world, how much faster could he have solved his many challenges?

Are all these sources of information and communication just a source of distraction? What with social media, streamed music and thousands of television channels full of entertainment, we are hardly left with any quality time to think!

In Leonardo's day, the process of publishing and distributing books did not exist. It took centuries for his own ideas and works to spread around the world. We take the ability to communicate very much for granted. It is this ability that has made possible the growth of Artificial Intelligence, robotics, and the manipulation of DNA. These are just some of the areas of science and technology in which Leonardo's original thinking has paved the way for today's innovations.

Today, we have a far greater community and social dimension to our lives, allowing us all to share our thinking and creativity. This benefits our lives, the organisations of which we are a party and our community as a whole. Tuscany was a tight-knit community and Leonardo was in the middle of it. However, that didn't hold him back.

Leonardo was a great traveller in his day, which was very unusual. Many people had to stay close to home to work on the lands belonging to their Feudal Lord. You could only travel with his permission. However, Leonardo managed to travel great distances in order to promote his skills and talents and find commissions for his work.

This was the Renaissance, and there was an appetite for new inventions, new buildings and new art. No one was better placed to deliver these than Leonardo. He saw an opportunity and took action. For him, action was always the key. He was never one to procrastinate. A good lesson to learn today!

However, despite his huge workload in delivering his commissions, he always found time to relax by studying nature and sketching the things he saw around him. A great way to relax the brain. Despite his workload, he led a balanced life.

One thing we share with Leonardo is our lifespan. He lived 67 years, a good age in those days! Even though we all live longer, thanks to medical advances and our understanding of lifestyle, it will never be enough! Leonardo was frustrated that he had not achieved more in his lifetime. He was very productive because he was 100% in charge of his time. Can we say the same of ourselves? A recent survey noted that, when at work, people spend around 12% of their time on social media. That is some 13 hours a week!

There are 168 hours in every week. Have you worked out where they all go? It is worth counting how much time you spend on everything. You might be surprised!
Sleeping–eight hours per night = 56 hours a week
Work–eight hours per day over five days = 40 hours a week
Travel–two hours per day over five days = 10 hours a week
A total of 106 hours.

This leaves you 62 hours in the week for eating, shopping, relaxing, hobbies, television and social media. Do you really need that time? Leonardo didn't, which is why he was so productive. What habits could you change to have more time? However, 62 hours is no longer accurate because of the time we now lose by using social media and apps. Be honest with yourself. How often do you break your concentration just to check what's happening? I dare you to total it. It might shock you!

I believe that today, we have a new Renaissance happening around us. Not everyone can see it, but it is definitely there. The dynamics and elements are very different from what Leonardo would have known. We have vastly more tools and resources at our disposal than in the 15th Century. So the question has to be—why are we not using them to their full advantage? Why are we not achieving more? Why are we not as creative as we could be?

Could the answer be that we are not 'hungry' enough? Are we trapped in our comfort zone? Is it the case that work is not just a four-letter word and not a motivation? Are we just downright lazy? We could spend an entire week in the armchair watching entertainment on demand on our televisions and surround sound speakers. Uber Eats will bring us food and drink when we need them, so there is no incentive to get up and cook.

No need to go out to meet friends; we can chat with them when we want to on our smartphones. The next thing will be to plumb us in so we don't even need to go to the bathroom! Maybe the film 'The Matrix' had it right!

Now, I am not encouraging you to go in this direction; I simply want to point out that we have far too many distractions that steal our time, steal our energy, and demotivate us from realising our true potential.

What I have in common with Leonardo is the wish to travel and see the world, but then, like him, I returned to Tuscany. Why? Because it is one of the very few regions on the planet where, for centuries, it has reached the optimal balance and harmony between nature and quality of life. The Tuscans have developed a culture and a tradition which, for me, represents true civilisation. Tuscany has a long, solid history and a formula that provides a good 'launching pad' for developing myself and my vision. Are you on the 'right pad'?

Having had the privilege of living and working here for many years. Looking at the rest of the world from this viewpoint, I see that evolution means that everything is amplified and subdivided into almost unlimited ramifications and variations. Everything you can think about already exists in some form or shape somewhere in the world. It is the opposite of what it was 500 years ago. Today, we live in a constant Renaissance where unexpected and respected new inventions can change humanity and our way of living forever!

If you want to be a modern Leonardo and achieve far more than the vast majority of people will ever think of trying, it is far easier to do so now than ever before. You have so many tools, platforms, networks, and information at your fingertips, and there is nothing you cannot achieve if you turn your mind to it. The opportunities are huge if you are curious, motivated and want to make your mark. You can be a modern Leonardo if you want to. Well, do you? (Warning: you might need to get up from the sofa!)

A NEW CITIZEN OF THE WORLD
Blending the real world with a virtual mirror of it, open to new dimensions, that we never thought existed, ready for all of us to use. Starting from our three dimensions: height, length and width, we will expand with the fourth one: Virtual and then, the fifth: Time.
New opportunities multiply endlessly.

He who is fixed to a star does not change his mind.
Leonardo da Vinci

A modern Colosseum, the power of combining shapes and functions.

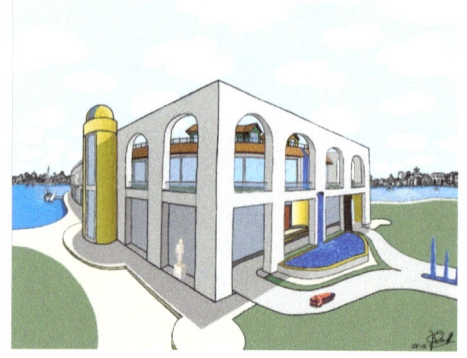

Design concepts from various centuries blended into one home.

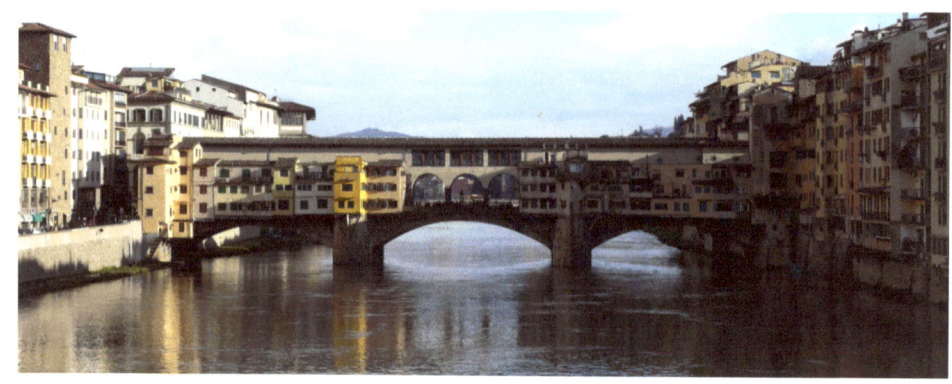

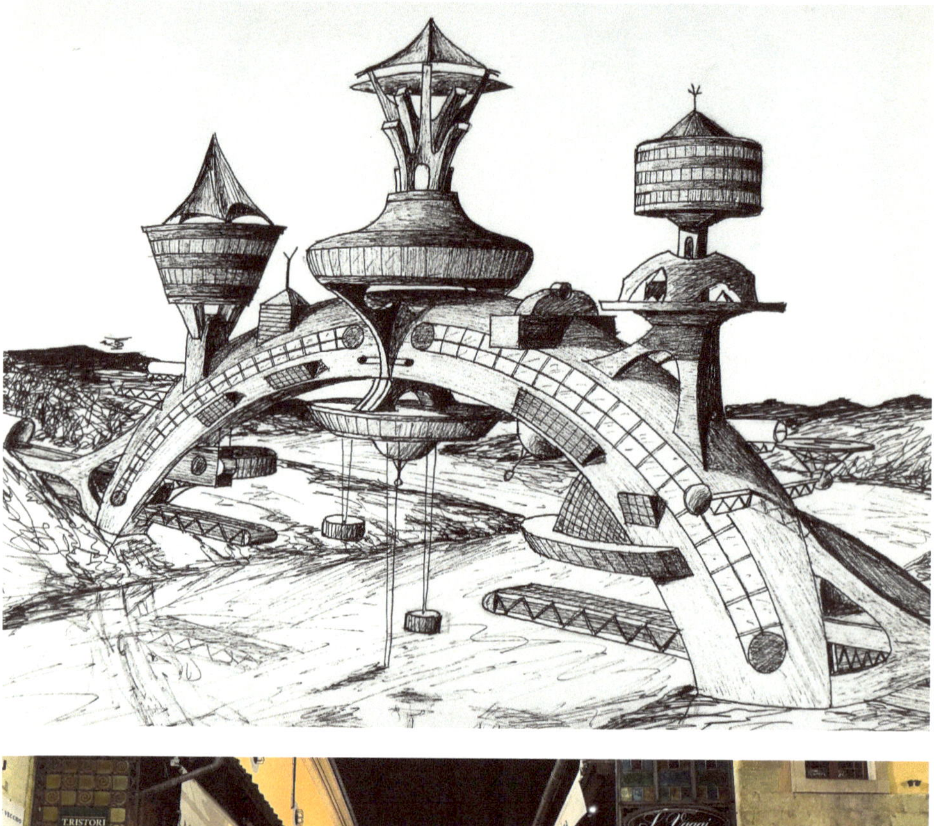

T.RISTORI
MARCO BICEGO
S. Vaggi
area
pedonale

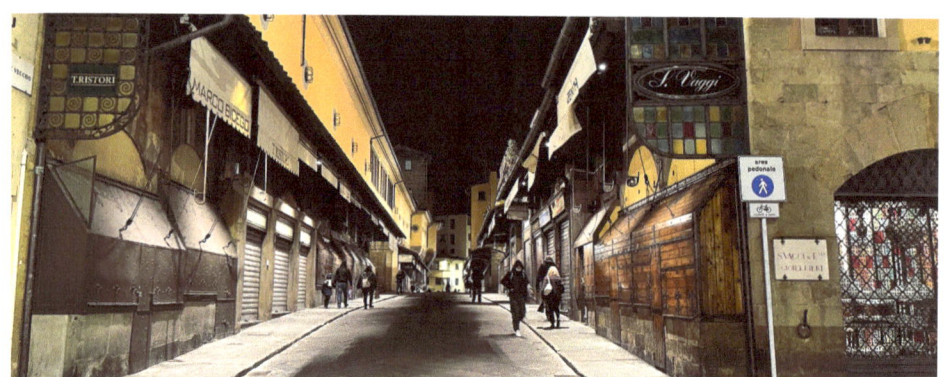

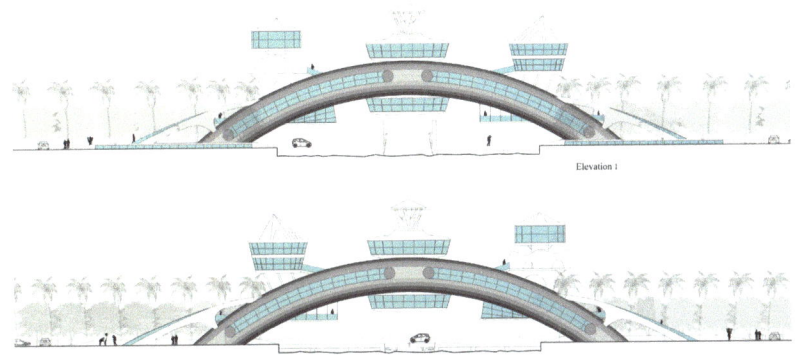

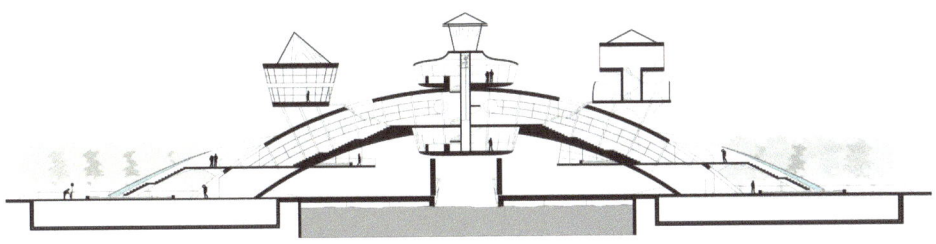

The Florence Bridge transformed into a New Futuristic Experience.

THE SPIRIT OF LEONARDO

He was the Master of Masters, and his life was devoted to understanding 'The All' and creating only 'The Best'. I was born in Rome to a Florentine family, with a vision of the World from long ago, driven by inspiration from the land of great artists and the Maestro Leonardo. Life is a great opportunity for doing your best.

Where the spirit does not work with the hand, there is no art.
Leonardo da Vinci

Chapter 5

Personal Connection with Leonardo da Vinci

The time has come for me to step out from behind my computer and properly introduce myself to you. As you can tell from the first few chapters, I am passionate about Leonardo da Vinci. My name is Avio Mattiozzi Petralia, and I was born in Rome into a family of artists and musicians. The earliest known of my artist ancestors was Rodolfo Mattiozzi, who was recorded as a composer and conductor in the mid-1800s in Florence. In the mid-1900s, my grandfather Tito Petralia was one of the well-known Italian conductors and composers. My late beloved brother Eliano Mattiozzi Petralia followed in the footsteps of our ancestors. So, music and art are firmly established in my DNA!

My early studies centred on art, architecture, and music in Florence and Rome. My father, Rodolfo Mattiozzi, was an active and accomplished painter and I exhibited for the first time my drawings next to his paintings when I was fourteen in a prestigious annual exhibition, 'The 100 Painters', in via Margutta in Rome.

My artistic career progressed, and by 1990, I was working internationally as an artist, not just in Italy but also in the UK, France, the United States, New Zealand and Indonesia. I continued my studies in fine arts, architecture, and interior design. I was enjoying the creative challenges of designing unusual and innovative art, villas, interiors and furniture.

My inspiration comes from everything around me, and I am particularly fascinated with technology and the materials from which many artistic manifestations are born. I have been using the precision of laser and water jet machines to create art using wood, marble, metals, plastics, plexiglass, leather, and fabrics. On the digital side, since the 1980s, I have been experimenting, developing and producing digital art, three dimensional and multimedia content. In recent years, with the use of immersive virtual reality, augmented reality and artificial intelligence, I found myself in an even more challenging environment.

In Leonardo's time, he could only work with the materials that were close at hand to him. Fortunately, I have something that Leonardo didn't have: the world of digital and computer graphics, plus the science of photography.

Leonardo once said, 'Art is never finished, only abandoned.' However, with the technology that we have today, it is possible for an artwork to evolve, respond to modification by the viewer and morph into another form of art. In this way, art can continue to surprise and delight for years.

Creative lighting can completely transform an object or a building. Art need not be the same each time it is viewed. In the South of London, there was an art project in Croydon, Surrey, called Skyline. With the help of a grant, many of the office blocks installed lighting and projection systems to bring vibrant colour changes into a bland concrete urban environment. A huge and dramatic art installation.

The Genius of the Painter

When the power of ruling Medici family ended, Anna Maria Luisa, the last Medici heiress, donated their extensive collection of paintings to the city of Florence. The collection is breathtaking, featuring works by Leonardo da Vinci, Michelangelo, Raphael, Titian, Caravaggio, and Rembrandt at the Uffizi Museum in Florence.

Studying their works for some time makes it possible to make 'a personal connection' with these magnificent artists. I spent enjoyable hours looking closely to observe Leonardo da Vinci's paintings, The Annunciation and The Adoration of the Magi. After a while, I began to think that Leonardo was by my side, inspiring me to push my creative boundaries and go beyond.

Leonardo was far more than just a painter, a scientist, an engineer or an inventor. He was on a lifetime's search to discover what linked everything together and the forces which caused everything to exist. He was looking for the door to bring us to the next level of knowledge and wisdom. His curiosity and capacity to do deep research into life's unanswered questions were insatiable. This is why I was attracted to his various capacities from an early age.

Painting was just one of the ways that he used to represent the minute details of the world around him. His paintings tell us far more than what is captured on the canvas. They also demonstrate how he studied nature and, indeed, his own body in order to discover how it was created. All this information is there for anybody to see, but Leonardo's powers of observation were on a completely different level.

His paintings also contained many levels of code. Nothing was included just by accident. There are hidden messages in his paintings that are just waiting to be decoded. That is why you need time to sit and appreciate them. He used many different languages, not just words but symbolism and metaphor. Every painting is full of multiple layers of meaning.

Leonardo added many realistic details to his paintings; he had so much knowledge of nature that he couldn't help but share it. If you look very closely at each of his paintings, you will be amazed at what you find hidden in plain sight! For him, every painting was a universe in its own right.

In the Universe, everything is in motion. The air, the water, colours, trees, people and animals are in motion. Leonardo wanted to capture this in a microsecond of a pause. As any video editor would know, a second is a long time. In fact, a second is made up of 24 'frames'. A blink of an eye equates to a twenty-fourth of a second. That is what Leonardo was trying to capture. If he were to look at the next 'frame' of his internal video, it would be different. Things had moved or changed.

Leonardo was reluctant to release his paintings as there was always something more he could add. Because he had studied nature in such depth, when he painted a person, he was able to give skin a translucency that revealed the shape of the body beneath, the bones and the organs. Incredible detail and depth of field.

My greatest insight into Leonardo's work was when I was involved with Centrica, a company specialised in developing cultural digital content and software, which was commissioned to make a very high-resolution scan of Leonardo's painting, The Annunciation, his earliest major work. It was painted in Florence between 1472 and 1476, when he was around 20 years old.

It was such a privilege to be able to work so closely with the great work. We used a top-of-the-range 100-megapixel camera, subdivided the painting into a grid of 8 sections and took close-up images of each of them. Then, we pieced them together into one huge digital image. The resolution was such that a speck of dust could fill the screen.

As we studied this huge image and zoomed in on the small details, we saw some small Spring flowers in the painting. As we zoomed in closer, we saw those small flowers had minute petals. My immediate question was, 'How did he do it?' There were no microscopes and no magnifying lenses in his time. Secondly, 'Why did he do it?' They were far too small to see. It was his desire for it to be the very as close to nature as possible. Amazing!

In the period when Leonardo was staying in Rome, invited by Pope Leone X, he decided to secretly work on something he had never revealed, not even to the German scientists who were producing sophisticated mirrors custom-made for his project. Some studies assumed that Leonardo was experimenting in a sort of photographic dark-room, and was involved in creating the famous image of the Shroud of Turin. Impossible? Well, for the Master Leonardo, certain things were possible.

Leonardo Through the Lens

There is no doubt that Leonardo had a photographic memory and an incredible ability to observe and to see minute details. Photography, as we know it today, didn't exist in Leonardo's time but his ability to create pictures in fine photographic detail still amazes us today.

Even though he didn't have a camera, he did what photographers do today by taking 'snaps' of the things they see. In his case, he would make quick sketches to refer to later. There are so many such sketches that, even now, a good number have yet to be seen or evaluated, and unfortunately, many were lost.

Taking advantage of our mobile phones today, we are able to capture in close-up details of those things that catch our eyes, and record them for later study as Leonardo did. Whether our interest is in landscapes, buildings, people, plants, birds, insects or mechanical constructions, developing such a library is following in the footsteps of Leonardo.

When he travelled, Leonardo would take with him a painting he was working on and bring it around, like nowadays, carrying an iPad. He worked on the Mona Lisa in this way, constantly making improvements wherever he went. It was as if he couldn't bear to stop working on it. The finished result we see today shows his desire to achieve perfection.

However, it wasn't just the image itself that fascinated him. He was looking deeper to discover 'the meaning of everything'. What was the divine purpose of everything? How did everything connect with everything else? What were the invisible sources that powered nature? He wanted to connect to the universe and tap into its energy and its secrets. He followed this mission all his life and never stopped learning and exploring.

I can remember, at the age of sixteen, being allowed to play with a Rolleiflex camera belonging to my father. This triggered my interest in photography, which became a significant part of my creative life as I grew up. As photographers, we are able to focus only on what the camera can see through its lens. Everything else is irrelevant. This is a great lesson to learn. Although Leonardo did not have a lens to help him, he trained his eyes to see even the smallest details.

I have followed in his footsteps as I have travelled to the places he grew up and those he travelled to visit. Seeing what he saw and where he obtained inspiration helped me understand what an incredible brain he had.

I have found that the software tools we have today can be used in far more creative ways than the manufacturers intended. All tools are only as good as the users and what their imagination can see. What the brain can imagine, we can find a way to bring it to life.

Today I have taken my personal interest in photography one step further and into the realms of augmented reality, integration with Artificial Intelligence and photogrammetry. Leonardo would have been so excited to have these visual tools to play with–but it has to be said, he did not do too badly without them!

Digital Art

Being a proper photographer requires a lot of study, practice and a large amount of specialist equipment. It has been a respected profession and a real art. However, the introduction of digital has changed all of that. Now, it is possible to take a picture on a smartphone with almost the same quality as an expensive camera. What is more, you can edit and manipulate the image on the phone and create the sort of sophisticated results that only an expert photographer could have managed in the past.

Because it is so easy to take pictures, people now take hundreds of them when in the past, they would have only taken a few. Our digital footprint has grown exponentially. When you add up the data from personal devices, social media, online activities, sports and exercise data, medical records, etc, the average person gathers several petabytes of data over the course of their lifetime. This is far more than we could ever properly curate, and so much ends up spread across multiple platforms and often ends up being lost.

In Leonardo's day, 'Less was More.' He may have sketched his subjects using multiple pieces of paper, but the number of his finished paintings was surprisingly few, around twenty. For him, it was quality over quantity. We are still marvelling over his work, 500 years later.

If Leonardo were to land here tomorrow, he would not fazed by the new technology because, in his time as now, for him, it would be just a means to an end, not the end itself. His focus would have been on creating the finest paintings and he would have used whatever materials he had close by to achieve it.

In 1977, I was an 'early adopter' and purchased my first computer with a massive 2 MB of RAM. Today, I am sure there is more of a memory in a musical Christmas card! This was the beginning of my love affair with computers and what they could do! It was the on-screen graphics that intrigued me. In the early days, there were very few graphics tools, so we had to create our own. With my two best friends, Pietro Galifi and Stefano Moretti, we set up our computer graphic company, Altair4.

When we started the company, we entered an unknown dimension at full speed. I completely believed that 'this dimension was the future' and the right way to go. Hindsight is wonderful now that the digital industry is global and well-established, but it was very different 45 years ago with no hardware, software, manuals or clients. There were not many people to share our vision with or even understand it. We were ahead of our time. All we could do was work hard to prove ourselves right.

We started with simple computer animations and creating three-dimensional objects. It was not just about proving that these tools and techniques were viable, but it was also about changing the mindset of how we used them. We could see that the status quo would continue and could visualise a very different future. This was the birth of new technologies and new processes. Not everybody could recognise the changes that were on the horizon.

We were alone in the marketplace at this time. There were no competitors, and people were struggling to understand the relevance of what we were doing. Leonardo must have felt the same way. He also worked alone, and few people had any clue what he was up to!

Forty-five years ago, I was at the very start of the digital industry. I don't think even Leonardo could have visualised how far it would have come today. With broadband, fibre, streaming services, home cinema, and thousands of television channels, who could have anticipated where it would go? And where will it be in the next forty years' time? We were not just thinking outside the box but had thrown the box away!

Nobody has a monopoly of boxes to think from. It is open to anyone to apply creative thinking to any new idea and to do the same in one of hundreds of different fields. The world is full of 'What if?' questions. What it lacks is sufficient people who are curious enough to find the answers to those questions.

The Hendrix of the Renaissance. Leonardo, the Pop Star

Leonardo's musical genius is not well known. When he was about 31 and a young artist in Florence, he was invited to the court of Milan—not to paint, but to play a customised *lira da braccio*, which looked like a fat violin, that he had made. It mostly consisted of silver and wood in the form of a horse's skull—so the harmony might be of greater volume and resonant in tone. However, there were not even written scores for his instrument. So he extemporised on it!

Playing it was 'the Renaissance version of playing the blues'. He used a friction belt to vibrate individual strings (similar to how a violin produces sounds), with the strings selected by pressing keys on a keyboard (similar to an organ). Leonardo took it with him on his travels and used it to improvise his own music. This was just one of many musical instruments invented or improved by Leonardo, among them various flutes, drums, hurdy gurdies, stringed instruments, and forerunners of the modern keyboard.

There is also strong evidence that he enjoyed composing poems and singing them with his customised *lira da braccio* as an accompaniment. The Hendrix of his time.

As a musician, I marvel at Leonardo's ability to express himself in this way and his ability to create instruments. Sadly, he left no music behind for us to explore further. For me, music is a great gift that has been in my family for years. Music embodies balance, harmony, elegance, power, mathematics, sophistication and beauty. It is empowering, energising, joyful and healing.

The *Viola Organista* that Leonardo invented produces a vibrant string sound that creates the impression that multiple instruments are being played. In some ways, it is like a small orchestra in its own right. No wonder he took it with him when he travelled. He would have been the life and soul of any party!

Raw Creativity!

I believe Leonardo's sheer energy and hunger for knowledge would have made him very challenging to keep up with. Few people today could match his output and the multiple topics that he was obsessed with. He gave everything he touched his full attention and focus. He was not restrained and had total freedom to go wherever his imagination took him. His attention to even the smallest details made him a true Master. It was as if time didn't mean anything to him. His hunger to complete a project or to discover something new would drive him forward.

*Details make perfection,
and perfection is not a detail.*
Leonardo da Vinci

Sitting in a state of mental freedom.

The History of my Life, yesterday, today, and to come.

THE VALUE OF LIFE

In simple ways, we can expand our interaction with ourselves and with everything around us. Many things have been already told and well analysed, but nevertheless, we are all individuals, different from each other, and we are passing through the eras of our lives with many different circumstances, factors, events.

As iron in disuse rusts,
so not doing, ruins the intellect.
Leonardo da Vinci

Chapter 6

Shifting Perspective, Take Actions

Leonardo's greatest legacy was not in his paintings, designs or inventions but in how his life and work inspired others to think in the way he did. I have tried putting myself in his shoes in many things I do. I have developed several innovative concepts and looked into them as he might have done. I hope this book will challenge your creativity and encourage you to do the same.

So much of our modern world would seem like magic to Leonardo, given how different it is from his knowledge and experience. Similarly, just fifty years ago the technology we have today would have felt like magic to us. Streaming video and the internet for example simply didn't exist back then.

As Homer said in ancient Greece, 'There is nothing new under the sun'; we are regularly surprised by the unexpected ingenuity of man. But after the initial surprise, we seemed to take a very long time to take up something new.

Back in 1934, the very first 3D film was released in France with the enticing title 'The Arrival of a Train'. Over the decades since then, the novelty of 3D films has come and gone, and today, it still hasn't become mainstream.

In 1900, the children's book 'The Wizard of Oz' was first made into a film. However, it wasn't till 1949, released by MGM, that new technology enabled it to scoop the Best Picture award. At a time when films were exclusively in black and white, it was no surprise to anyone that the film opened on a small screen in monochrome. At the point where the lead character, Dorothy, lands in the fantasy land of Oz, suddenly the film switches to glorious technicolour and simultaneously jumps to a huge panoramic cinemascope format. At the time, the impact of this was stunning and triggered spontaneous applause. So, how could we look at things in our world in a different way and perspective creating something entirely new?

Today, colour is no longer unexpected; in fact, it is our expectation. We no longer have large cinema screens; instead, we have our home cinema systems on big flat-screen televisions. These, too, are no longer a novelty, and we take them for granted. We seem to have completely forgotten the old Cathode Ray Tubes (CRT) that used to be on every television. So, where is technology taking us? As we fully immerse ourselves in the world of digital imagery, new technologies have emerged.

ABBA Voyage is a virtual concert residency by the Swedish pop group ABBA. The concerts feature virtual avatars (dubbed 'ABBAtars'), depicting the group as they appeared in 1979, and utilise the songs' re-recorded vocals from the group themselves in a studio in Sweden specifically for this show, accompanied by a live instrumental band on stage. The dancing and singing are all captured from the studio set, so it is all of them except for their younger Computer Generated Imagery (CGI) avatar appearances. Unlike earlier digital avatar performances (sometimes referred to as 'hologram' concerts), ABBA Voyage plays out on 65-million-pixel LED screens. In previous shows featuring the likes of Roy Orbison and Whitney Houston, performers' avatars were projected onto a band of translucent plastic.

Despite the amazing level of technology now available, our experience in browsing websites online is at least a couple of decades behind what is actually possible. The internet is still in the dark ages. Everything is a flat two-dimensional experience. Site navigation is still using menus and buttons. Very few websites take advantage of 3D graphics despite 3D having been around for more than 100 years. The internet today is a flat and predictable world with only basic interactivity. It is time to drag our browsing activity out of the 'middle ages' and into the multi-dimensional world that has existed for quite some time!

This is now about merging the real world with the digital one, bringing the two together in new viewing devices–not restricted to a fixed 2D flat screen. The vision presented in the film The Matrix was that we would be able to interact with one another, move in any direction, use digital objects and conduct meetings and business within a digital space. This reality should be available now and part of our daily lives, as the necessary technology exists! There is no turning back!

Now is the time to choose your Avatar! How will your personal Avatar be dressed? How will it represent you? Will you insist that you look younger than you actually are? Will people be shocked to see what you actually look like in the real world after meeting your Avatar online?

Will it be a bit like Oscar Wilde's 'The Picture of Dorian Grey', with your old self captured in a painting, leaving your avatar to represent you online? Will we send out our digital alter-egos to go and collect the digital shopping and have it delivered to our homes? There is so much ahead of us to look forward to!

Will we send out our real humanoids or virtual avatars on holiday and to explore the world on our behalf? With the amount of trouble and unrest in the world, it is not as silly as it sounds. I have been working for several years on a concept that mirrors the real world in a virtual one. Not only can everything be visited and operated in the real dimension, but also in the virtual parallel one, as these two worlds are fully interconnected. This will introduce additional dimensions beyond the three we already experience in our real environment, creating an exponential number of new opportunities in how we live and operate today. Furthermore, it will provide new and immersive experiences from the comfort and safety of our own homes! I elaborate more about this concept on pages 204 to 207 and in Concept 7 inside the website: The Spirit of Leonardo. Both can be accessed using the two QR codes provided below.

Let me ask a question. If Leonardo had access to all the tools, technologies, and facilities that we take for granted today, what would he do? What insights would he have that we might overlook? What connections would he make that have yet to occur to us? What would inspire him the most?

Leonardo was drawn to nature and the physical world in which we live. He was also drawn to the human world with all its complexities. But, in his time, the digital world did not exist. It is hard to believe that the internet as we know it today only started a few decades ago, in January 1983.

How would Leonardo have embraced this amazing world? He would be surprised that, after an initial twenty-year burst of development in the 1960s, we have allowed ourselves to settle with a two-dimensional screen format when there are many innovative ways to search and access information. We have restricted ourselves to look at the world through small 'windows' when we should have removed the wall altogether.

The world is in three dimensions, but we continue to try to access it with the old two-dimensional technology and a 'mouse'. I don't think Leonardo would have allowed himself to be restricted that way! He would see the need for a more intuitive interface that allows us to access what we are looking for unconstrained by dimensions.

While today's virtual reality headsets and similar devices allow us to enter and immerse ourselves in digital spaces, there is still a long way to go. I have been studying and designing three-dimensional environments that people will eventually inhabit. However, in order to reach that point, we can currently only use two-dimensional tools. Would Leonardo see something we are missing? What would Leonardo do? He would create something new. What would you do?

Leonardo the Artist

Everything that Leonardo did he first examined in detail before starting. He searched his brain for similar projects he had worked on and searched through his treasure trove of drawings and notes he had created over the years. But more than anything, he allowed his brain the space and the quiet to be completely creative. This was his system, which enabled him to be fast and efficient in coming up with new concepts. Let us explore his processes and the thinking he applied to each concept–The Leonardo Way.

Leonardo himself was very private in the making of his creations. He would work in secrecy. When he used other craftsmen to help him with a project, he often swore them to secrecy. He only shared with them the minimum information required to do their job, never the big picture.

All the great artists of the time, like Andrea del Verrocchio, the Pollaiolo brothers and the Della Robbia started their careers as independent artists. Still, as they grew in skill and fame, they created their own *bottega dell'arte*, which would attract young artists who became their apprentices. The leading artist who ran each *bottega* was famous enough to attract commissions from wealthy noblemen. However, they had more work than they could handle by themselves. Their apprentices would share the load and could fill in details if needed. Craftsmen in the group would make and prepare canvasses and mix paints. It became a creative production line where everyone had a role. By being a part of a *bottega dell'arte*, artists and craftsmen had the opportunity to learn their craft from a master. They didn't have to worry about searching for new commissions or marketing themselves.

Throughout the Renaissance period, by joining together into *bottegas*, individual craftsmen were able to attract commissions from wealthy merchants and noblemen looking to impress their fellow countrymen with their art collections. By banding together, the artists were able to influence and inspire each other and quickly find fellow artists to help them deliver a large commission. This quickly became a large and profitable industry which had not existed before. It was an outburst of amazing creativity that became one of the hallmarks of the Renaissance period.

By contrast, Leonardo was more of a loner. He never had his own *bottega*. He made a name for himself by offering a wide range of solutions to the princes and nobles he served. In addition to his ability to create impressive art, he also had a talent for creating practical inventions, weapons of war, and machines, giving him an edge over his rivals.

He recognised that wealthy nobles and merchants wanted to show their wealth to their friends. There was no better way than to fill their homes and palaces with the best entertainment, engineering solutions, and inventions. They were all trying to outdo each other. It was a sellers-market, and Leonardo knew how to sell! He also knew what it would take to turn this into a business. Leonardo never thought small! The scale of some of Leonardo's inventions was impressive. His clients were the most famous people of their day; among them were the Medici Family, Ludovico Sforza, Pope Leone X and the King of France.

If you were to compare the *bottegas*–the collective art workshops of the 14th Century with today's art, which do you think is the most important? In Leonardo's time, there was an insatiable demand for art in all its forms. This was the era of the great painters and the old masters. Their work is still exhibited today and is now worth millions. However, compared to today, the number of people working in the creative industries in the UK alone is 2.3 million. They generate a staggering £108 billion to the UK economy.

If you think about it, art is everywhere. Every website needs to be designed. Every building features interior design. Design is featured on packaging and publications. Art is featured in every city centre's design and public space. If we broaden our definition of art from simply those paintings hung in a frame and include the vast arena of digital art and visual entertainment, thousands of designers work daily in this huge industry. Art is art; only the media and the settings have changed. It is still all art.

Design shapes the world around us and creates opportunities for creative talent to flourish. Many careers can be based on art. Today, art is made up of many individual artistic enterprises, just like in the art workshops (*bottegas*). There is a thriving industry where everyone can become an artist in many different media. Art itself can be displayed in many different ways, further enhancing the viewer's end result.

There are some towns, like Folkestone in Kent that focus entirely on art. There are galleries, shared studios and work spaces designed to attract artists to work there. In turn, these draw in visitors to see and purchase what the artists create in all its many forms—A modern-day *bottega*.

Few professions offer diverse opportunities to build a business based solely on your imagination. This opportunity is open to everyone with the necessary talent.

Leonardo challenged people to look at art in new ways, to see it as a number of infinitely interchangeable elements, not just a single canvas. Art today can be even tactile, touched, stroked, manipulated, and interpreted in unique ways. Images could be reversed, inverted, or rotated and see its juxtaposition with other images to create several different effects.

Now, not only would the artist be the one to interpret a piece, but also the viewer could adapt the way the art could be composed and displayed. They could create their own art using modular elements. You can see these concepts in today's art, which is made out of new and original materials. Artistic elements can be made with many materials, such as leather, wood, metal, plastic, carbon fibre, and more. Will Leonardo like to play with such an interactive formula?

The Cities We Live In

I became interested in the designs of towns and villages near where I lived. Later, this interest sparked my training as an architect. We all need and appreciate living and working in spaces that have been created from the human imagination and heart.

Every town or city has a soul and an identity of its own. No two are the same. In every case, a city's charm was no accident but the work of town planners, innovative builders, and retailers vying for attention, surrounded by architecture that may well have been there for centuries. This unique combination is what created the magic.

Every city has its own unique 'style of being', blending together its past and its present. Each city has its own story to tell that projects happiness, merriment, sadness, chaos, bustling, abandoned, or crowded.

Some cities are home to millions of people and possess distinct characters. However, every city is unique with its own identity, its soul, its colours, the sky, and its own atmosphere. It is a live entity and a source of energy and fascination to everyone it welcomes.

Living in one part of a city and having to commute on public transport for an hour in either direction can be stressful. It is equally stressful to work in an office or factory all day. Life can be challenging. So when a city recognises this and realises the need to improve the environment through designing inspiring public spaces, a huge difference can be made in the quality of life. Green spaces give a city space to breathe and de-stress. Much creativity has often been applied to help the old part of a town integrate with new constructions to great effect. Clever design is always a source of pleasure for the eyes.

What would Leonardo's unpredictable mind propose today in order to improve the way our cities operate and their aesthetic look? Do you have some ideas in mind?

Collaborative Art

I believe in the concept of collaborative art, where the artist and the viewer collaborate to create the work. A simple way to see this is with fashion. The designer might create separate items such as gloves, hats, skirts, shoes, scarves, and even jewellery. However, it is the person who wears them in different combinations to create several different looks.

Art does not have to be a static or unchanging image; it can evolve to suit the viewer's mood and imagination. Art does not just have to be on canvas; it could be generated by light and projected onto a surface. It can even be made to morph using polarisation. A well-designed artwork can create unlimited variations.

With the use of different angles and lenses, impossible perspectives can be created. Think of a Fairground Mirror! For even more outrageous images, photo manipulation software can build selective layers into a completely unique result.

In studying Leonardo over the years, one thing stands out. His energy! He was on a mission to understand everything. What are things made of? How are they assembled? How do they work? And more importantly, for him, how do they link to the Universe as a whole? How are they part of the great divine plan?

For Leonardo, every day was a source of wonder and discovery, and I believe that those around him would have been caught up with his creative enthusiasm. From him, I have learned to keep opening my mind and see more possibilities in every situation.

To adopt this mindset, get into the habit of asking more questions to yourself as you observe. Everything around us has a story to tell. A story about its origin, its past, its role today and its value in the future. Thinking the Leonardo way! He captured every thought and idea on paper. He was massively prolific with his notes, which he stored and was able to refer back to.

In today's digital world, we face a growing challenge. Our archives of notes, thoughts, images, recordings, and documents are no longer in physical form. They are data. As such, they are vulnerable to loss because of technical glitches and hardware failures. A life's work can be inadvertently lost at the click of a mouse. All of the cloud storage in the world can be for nothing if it is erased by rogue data.

A cautionary tale happened in late 2023 when the British Library suffered a serious attack on its computer system by hackers. It closed down the organization for months as they had to do a forensic sweep of all their servers. It was something that couldn't happen, but it did.

Now, hardly a week goes by without organisations and public bodies reporting data loss from a hacking attack. International terrorism can be achieved simply by disrupting the internet, attacking satellites and cutting undersea cables.

The answer to all of this is not to revert to storing boxes of handwritten notes! However, some new thinking is needed to create different strategies for the future. A new vision for a more dangerous world.

Leonardo was a one-man research laboratory. His legacy of notes, the results of his studies, his research, simulations, experiments, and his discoveries of possible procedures filled his days.

The very act of writing something down on paper creates a powerful link in the brain. A memory. If that memory is made strong enough using memory techniques, the human brain is capable of retaining vast amounts of information.

In 1991, Tony Buzan, author of multiple books on the brain and also the inventor of Mind Mapping, had a question. Why were there many competitions involving throwing, kicking, or catching a ball but no competitions about using memory? When you think about it, you realise that we need memory for absolutely everything we do. No memory? No language, no recognition of other people. No ability to know where we are and where we are going.

In conjunction with Chess Grandmaster Raymond Keene OBE, Tony Buzan founded the first World Memory Championships. It was front-page news worldwide. The first winner, who subsequently went on to win it a further eight times, was Dominic O'Brien.

The competition was not about general knowledge or about how many facts you could cram into your head in advance. Instead, it was about how much new information you could recall in a given period of time and then recall it accurately against the clock.

Using simple techniques to make the information memorise 'brain friendly', the competitor would go on to recall the order of 30 randomly shuffled packs of cards in an hour, 4,000 binary digits in the same time, plus similar feats from spoken numbers, random words, and hour numbers.

My point is that, even a century ago, the average person had a vastly better memory than we have today. Over that period, we have come to rely on phones and gadgets to recall even the smallest of information. We have replaced our brains with machines. No wonder we are now so vulnerable to losing our precious memories.

This takes us back to Leonardo. The sheer amount of information he kept in his head would seem impossible to us today. And yet, the techniques used in the World Memory Championships prove that we still have the skill, if only we knew how to use it. Sadly, though, as the attack on the British Library proves, we cannot rely on machines to preserve our memories in the future. So don't throw away your pen and notebook just yet!

Leonardo, after his death in 1519, was considered a very important painter, but nothing more than that. This was because all his manuscripts were neglected and abandoned. Some had been split into parts. Much of his incredible and valuable work had been lost forever. Just in the late 18th century, his codes and notes were collected and restored. This portrayed Leonardo for the first time as the real multi-skilled artist, scientist, engineer, and visionary genius he was. If he had left it on a hard drive, we would never have known! A lesson for us all!

From 1849 to 2015, In the House of Commons, the British Parliament kept record copies of public Acts printed on vellum, a durable material made of calfskin. Until 1849, they were handwritten on parchment rolls (usually made from goatskin). Not a computer in sight!

The world today is moving at an astonishing pace. Just look at the speed at which Artificial Intelligence just appeared out of nowhere! The interesting thing is that each one of us has access to all the world's knowledge in a fraction of a second using a search engine. We have all of the answers! What we lack are the right questions. Leonardo was brilliant at asking the right questions. His genius was in seeking out the answers.

So, with everything now at our disposal, what excuse do we have for our lack of creativity and inventiveness? Why are we all not geniuses? Maybe it is not about knowledge but more about our mindset in how we generate questions?

Learning never exhausts the mind.
Leonardo da Vinci

Tuscany harmony is thoughtfully located in New Zealand.

The evolution of harmony between humans and nature in Tuscany.

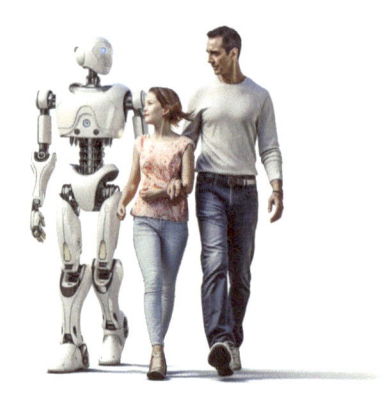
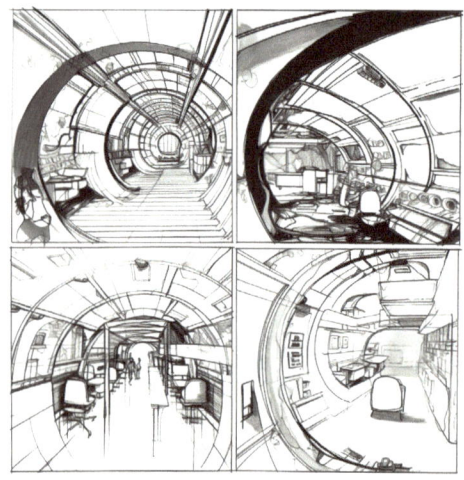
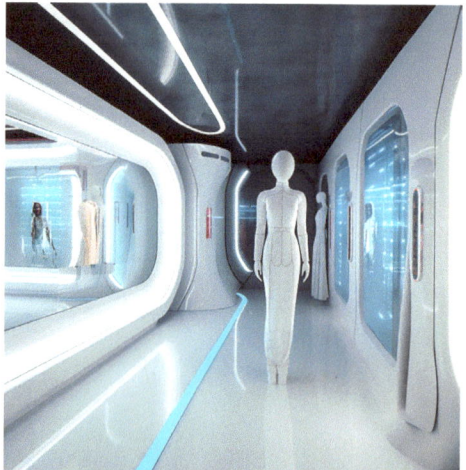

What else can we invent in the next one hundred years?

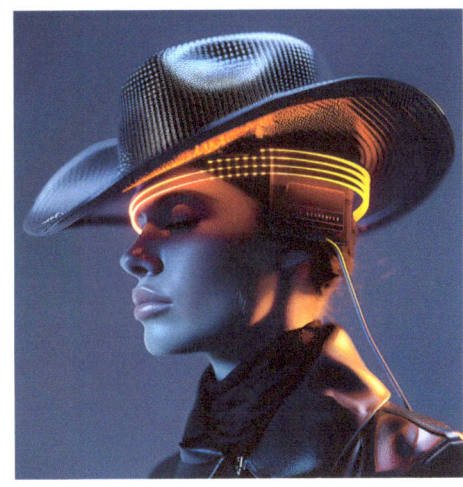

So many new things and experiences can add quality to our life.

We have accomplished so much over the past millenniums.

The best has to come with things that now we can not even imagine.

London in the real world, and in the parallel virtual world.

London in the past, present, and future, all in one dimension.

A NEW WAY OF LIFE

New tools, dimensions, worlds designed by different providers, working concepts, business models, and interactivity with everyone and everything close and far. All of this will radically change the way we all live in a progressively much more dynamic way. All is possible.

Those who have experienced flight will walk on earth, looking at the sky, because there they been and there they wants to return.
Leonardo da Vinci

Chapter 7

Curiosity, the World of 'What If?'

All of us have far greater capabilities than we could possibly believe. Our brains have infinite creativity and inventiveness. We do not require a 'label', a job title or even an 'ology' or qualification to look at something and to say, 'It's great, but what if...?'

Leonardo made it fashionable to be curious and to think, not just outside the box, but without boxes entirely! His meticulous mind needed to understand things in all their details. His constant speculation about 'What if?' drove him to imagine and create fantastical objects.

For the King of France, he created a mechanical lion that would walk in front of the King, and then a hatchway on its chest would spring open and scatter lilies at his feet to the delight of his court. Leonardo knew how to impress and delight his benefactors and sponsors. This was the high-tech of its day and incredible to see. It was the forerunner to today's Artificial Intelligence.

The whole area of Artificial Intelligence originated in this early quest for developing machines that could operate on their own and be automatic. It has taken many centuries for this to become an overnight success! Leonardo was there at the very beginning.

This concept actually goes back thousands of years. Leonardo wasn't the first to create animated mechanical objects. One of the earliest records of an automaton comes from 400 BCE and refers to a mechanical pigeon created by a friend of the philosopher Plato. 'Automatons', as they became known, came from ancient Greek and meant 'acting of one's own will' and was the origin of the modern word automation.

Leonardo da Vinci wrote extensively about automatons, and his personal notebooks are littered with ideas for mechanical creations ranging from a hydraulic water clock to a robotic lion. Perhaps most extraordinary is his plan for an artificial man in the form of an armoured Germanic knight.

According to Da Vinci's sketches of the key components, the knight was to be powered by an external mechanical crank and use cables and pulleys to sit, stand, turn its head, cross its arms and lift its metal visor. While no complete drawings of the automaton exist today, evidence suggests that Da Vinci may have built a prototype in 1495 while working under the patronage of the Duke of Milan. In 2002, NASA roboticist Mark Rosheim used Da Vinci's scattered notes and sketches to see if he could create his own version of the 15th-century automaton. The Rosheim knight proved fully functional, suggesting that da Vinci may very well have been a robotics pioneer.

Creating practical mechanical objects was his passion. Because of the sheer volume of sketches that he made, people have been able to try building them and find out if they actually work.

However, Leonardo's greatest curiosity was nature and how the human body worked. His desire to really understand it drove him to conduct an autopsy on more than 30 bodies. As a result, his anatomical studies were incredibly accurate. His understanding of the different layers of muscles, blood vessels and organs means that his paintings of people were astonishingly lifelike.

Similarly, the whole field of botany made Leonardo curious. His sketches and paintings demonstrate his deep knowledge of plants, which he often shared in his paintings. The backgrounds behind his portraits often contain many recognisable different species of plants, flowers, and trees.

So what can we learn today by seeing where Leonardo's curiosity took him? Of course, we cannot all be polymaths; few of us would be capable of thinking like Leonardo, but what we can all do is look around and say, 'What if?' and let that trigger our creativity and imagination.

Did you make the decision years ago to stop studying and start working? 'What if?' you did both? 'What if?' you looked at developing your passions with an Open University degree and still following a profession. Where could that have taken you today?

You Never Stop Learning; You Only Stop Being Curious

Today, we have one thing that Leonardo didn't have: the internet. For the curious mind, this amazing resource allows us today to interrogate the output of the world's greatest brains. No matter what fact or knowledge we are looking for, Mr Google will find it for us in a fraction of a second.

Now, you can take your 'What ifs' and your questions to the world's best source of information. It is truly incredible that the first internet search engine wasn't launched till 1990. How did we manage before that? But there is a downside to everything.

The first known spam email sent was in 1978. Gary Thuerk, a marketing manager, sent approximately 400 people an email advertising his company's new computer model. His spam efforts reportedly earned the company $13 million in sales.

The next huge step forward was the invention of apps. Mobile applications are software designed to run on a smartphone, computer, tablet or other electronic devices. Apps have a design intended for a specific function. Most apps relate to a business or a service. Some relate to education or entertainment, which includes the world of gaming.

We now have so many powerful tools at our fingertips; what excuses could we possibly have for not rivalling Leonardo's outstanding creativity and output? This proves that technology is a red herring. Do we really need a bank of screens on our desks in order to change the world? Leonardo didn't! All we need are our brains and strong will to achieve anything!

Today, our fixation is all about speed: high-speed trains, supersonic flights, fast food, streamed video, fast cars. We are always connected 24/7 and trapped with the urgency of doing and knowing things. There is much to be said for the peaceful rural existence that Leonardo enjoyed.

In the year 2024, the Lume Melbourne in Australia organised for the first time an immersive digital exhibition using cutting-edge technology, 'Leonardo da Vinci - 500 Years of Genius.' Some pages from the Codex Atlanticus, the most extensive single collection of Leonardo's original sketches, designs, writings, innovations, and inventions on diverse subjects, were exhibited. As well as his world well-known painting, the Mona Lisa, was shown in three-dimensional with the use of a 240 million-pixel multi-spectral camera, revealing her famous expression. The exhibition also featured three-dimensional renderings of Leonardo's inventions and innovations.

The use of advanced technology was the central of the exhibition, showing that Leonardo's works have always been and will always be an inspiration and a trigger for innovation.

Leonardo is the best of geniuses, the variety and sophistication of his work, the complexity of his execution and his techniques are still the best today we can use for demonstrating and showing cutting-edge technology. The Lume Melbourne exhibition is a perfect example.

The more technology advances, the more we can enter into Leonardo's paintings, drawings, annotations, and the still unexplained ideas he wrote in his Notebooks. Jonathan Swift once said, 'Vision is the art of seeing what is invisible to others.' This was certainly the case with Leonardo.

No who begins, but who perseveres.
Leonardo da Vinci

On the right page: The time machine we all are, 20 years in London.

The blending of human capacity into natural perfection.

A portable home versus a futuristic, immovable fortress.

Creating new interactions between human thought and nature.

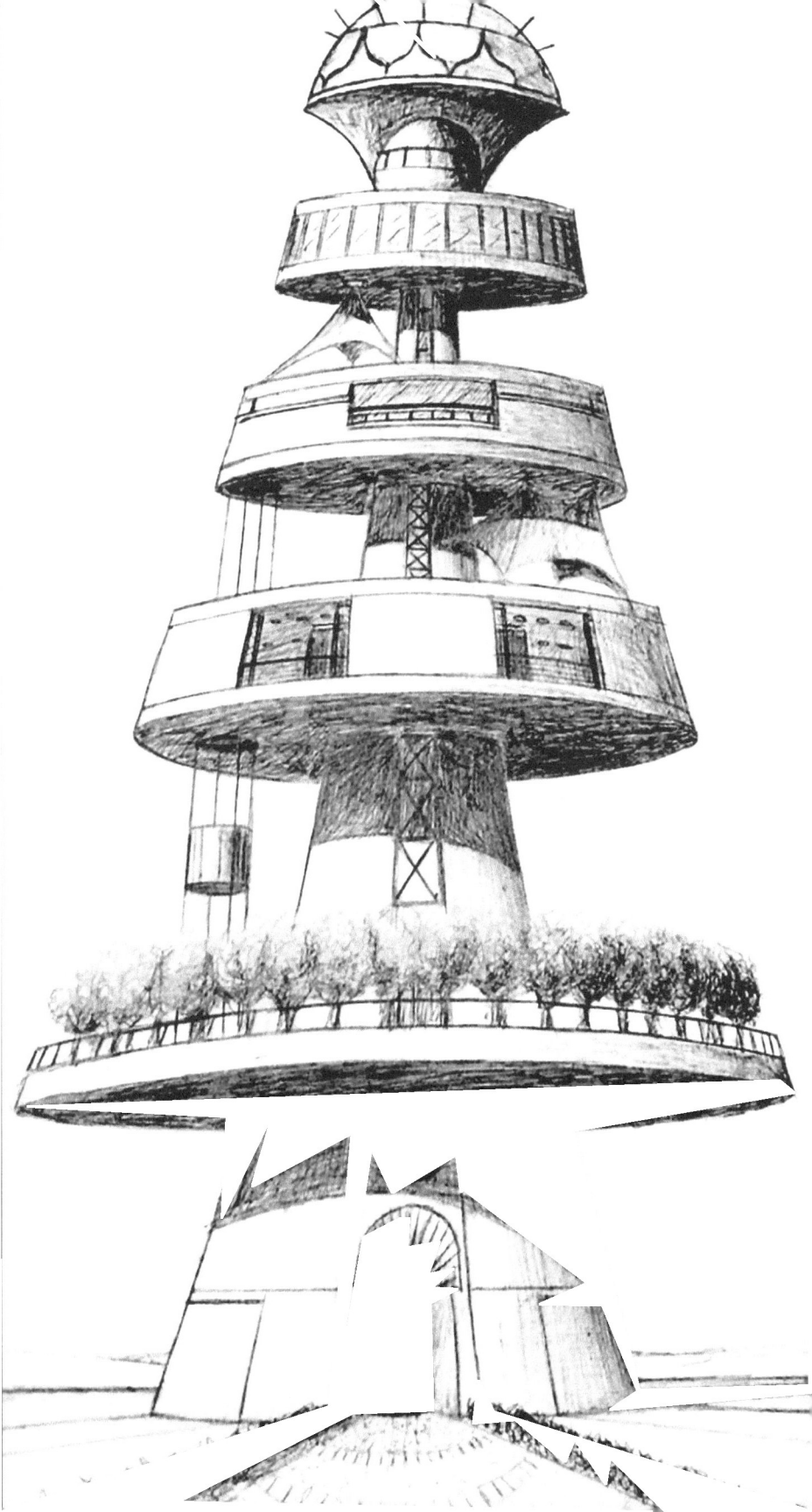

Flying homes can provide a completely new sense of freedom.

OUR NEW HOMES

The evolution of us all passes through the evolution of the way we live, of our lifestyle, and with so, of the place we live in. Our home should, even more than in previous centuries, represent the physical, intellectual, and practical environment that represents us and allows us to use it in multiple ways and for various tasks on the planet Earth and beyond.

In rivers, the water that you touch is the last of what has passed and the first of that which comes; so with present time.
Leonardo da Vinci

The noblest pleasure is the joy of understanding.
Leonardo da Vinci

Chapter 8

The Power of Observation

All of us have brains that are thirsty for understanding the world around us, the people we meet, the circumstances we find ourselves in, and the environment that sustains us. To do so, we need to employ our five senses: vision, touch, smell, taste, and hearing. Everything we learn comes to us as a combination of these and then needs to be deciphered and understood by our brains.

Many people believe that there is also a sixth sense—a supposed intuitive faculty that gives awareness that is not explicable in terms of normal perception. For example, you might sense that you are not alone in a room. In addition, a sixth sense could be proprioception, the perception of body position, which is important for balance and agility in movement. It could also include perceiving stimuli from within the body, such as pain, hunger, or thirst.

The function of the brain is to make sense of all the information that is constantly being fed to us through our senses and to reference it to what the brain already knows. It's an incredible feat when you think about it! With all of our digital features, including Artificial Intelligence, we have yet to create any system that can match exactly what the brain can do.

We rely on our brains to identify, sort and categorise the cascade of information we receive every moment of each day. How well the brain does this determines how effectively we function. For example, suppose you were at an overseas conference surrounded by a large group of people all speaking their own languages, none of which you knew. Suddenly, in the middle of this blend of unharmonious sounds, your brain picks up the one voice speaking in English. It flags this up as 'urgent and important' and brings it to your attention. However, there has to be something else at play, or we would all be responding like clones in an identical way to the same information.

At the same conference, imagine that, at the end of the event, you asked a sample of people what was the most important thing they learned. The chances are that each person would identify something different. Why is that? Because they were all there looking for something different in the first place. When their brains heard the one thing that their subconscious mind was looking for, it was immediately flagged up.

Leonardo's huge number of sketches, doodles, diagrams, and detailed pictures of nature show his inexhaustible quest to understand his world but also to invent and create a new one. He always asked questions but was secretive about his work. He would protect his ideas using mirror writing and codes. It is quite possible that his most creative ideas have yet to be deciphered from the mountain of his notes and sketches that are still waiting to be discovered. He also used a secret language. His world was dominated by Italian and Latin, but he also knew some Greek and French.

Leonardo added a lot of symbolism to his paintings, so the meanings were not entirely clear to the casual observer. There are visual references that he included that would mean little to us today but were more meaningful to the people viewing them at that time.

For a painting as famous as the Mona Lisa has become, many artists have felt the urge to paint their versions. There are many different versions; some have her with a moustache, glasses, or even a cigar. I created my own version where she is engraved onto transparent plexiglass. In normal light, she is invisible, and all you can see is her shadow on the wall behind her. However, when the lights built into the frame are switched on, the image comes to life in many colours.

Leonardo had a great advantage over the rest of us living in the 21st Century in that he had far fewer distractions in his life–and no technology! 500 years ago, it was much easier to have peace and quiet to observe and think. In fact, any creative process that needs our brain's active involvement is only possible when distractions have been removed.

The power of observation is a talent born out of curiosity that predisposes the mind to observe with more concentration. Good observation, then, is less about the eyes themselves, assuming that they are working correctly, and more about how our brain directs them to see deeper than we would normally do. Without a specific purpose to learn more about something, we tend to glance rather than look, skim instead of study.

By studying Leonardo's paintings, you cannot fail to be impressed at the deep level of detail that they contain. The closer you look, the more detail there is to see.

An example of his obsession for detail can be found in his Codex Leicester, which is a collection of his scientific writings. It is a leather-bound notebook containing 36 handwritten sheets of words and drawings. When the book was last sold in New York in 1994 for US$ 30,802,500, purchased by Bill Gates, it became the most expensive manuscript ever sold.

The Codex was written in his native Italian, but in reverse, as mirror writing. Today, Microsoft displays extracts of it as one of its screen savers. They have to be the most expensive screen savers in the world!

Let us continue with Leonardo's power of observation. Here's a list of some subjects that captivated him, which he diligently observed and documented throughout his life.

Δ The golden ratio / divine proportion theory.
Δ The relations and proportions of the human body parts.
Δ The working of the human body in relation to the universe's movement.
Δ The functionalities of the human body and all anatomical studies.
Δ The study of human anatomy led to the design of an automaton, also called Leonardo's robot.
Δ The study of nature in all of its glory, from plants, animals, birds, and, of course, humans.
Δ The reason sea creature fossils could be found in mountains.
Δ The flow of water in rivers and how it was affected by different obstacles.
Δ Effects of water erosion.
Δ The acoustics of churches and the capacity for voices to reach every part of the building.

- The luminosity of the Moon and why it reflected the Sun's light.
- Why the Moon was not as bright as the Sun.
- The reason planets shine.
- The study of biomechanics.
- Botanical study of plants and flowers, focusing on their growth and differences.
- Geology studies of the Arno River Valley in the Tuscany Region, as well as Milan and the Lombardy Region, including notations on weather conditions, effects, and changes.
- Accurate cartography of the city of Imola and Chiana Valley in Tuscany and other areas, using his own method, resulted in quite high accuracy compared to other cartographers.
- Study water motion in detail, including different forms and motions of water on various material surfaces, and investigate the spiral as it occurs in water.
- The experiments and ideas about natural fluid flow which were innovative for his time.
- The study of flying animals to be then applied to humans.

Certainly, Leonardo utilised all his observations, studies and research in everything he created, designed and invented.
In his paintings, for example, all the background plants, flowers, and botanical details would be easily recognisable by a botanist. Then, of course, all his accumulated knowledge merged together in his inventions, such as the parachute, the helicopter, tanks, the anemometer, and specific chemical formulas in order to protect and maintain his painting in almost perfect conditions up to today. Remember that this is the work of one man with no computer, camera, video camera, electricity, recording machine of any type, car or any sort of motor, internet or Google.

As a French friend of mine told me years ago, it must be the Tuscany air, the perfect harmony this region has always been able to achieve a relationship between nature and human presence. Perhaps, I can't say for sure, but certainly, I have to admit that this crucial capacity of observing has been the 'main motor' that moved me from my youth up to now in looking into anything, trying to understand how all works. Thinking about what can we change to make things better and more efficient, and how differently we can do things?

Today, after more than 500 years, every profession and industry has evolved significantly. Skills have become highly sophisticated, and almost everything around us has been thoroughly studied and analysed. In Leonardo's time, for example, the options for building materials and interior decoration were limited. However, in modern architecture and interior design, everything is complex, with precise regulations, engineering specifications, numerous materials, systems, lighting, and automation to consider. There are numerous styles to draw inspiration from and many manufacturers with extensive catalogues and specifications to study.

Leonardo's remarkable observation skills remain significant even today. He could see beyond the obvious, often noticing things that others overlooked, with a clear purpose and a desire to explore further. Leonardo and other geniuses of the past 500 years have been a constant source of inspiration and a role model.

Researchers concluded that one of Leonardo's superior strengths was his exceptional eyesight, which certainly intrigued and attracted me to his work because of my own very limited eyesight.

His work felt like an extension of my own vision capacity in some way. Synchronising his keen observation with my imagination allowed me to expand my ability to see beyond the visible things.

So, my approach involved studying his codex, notebooks, sketches, and paintings. I would spend hours in front of them, trying to 'communicate' with him and sense his presence and message, both obvious and not so obvious. I aimed to understand what he was thinking when facing something and how he was attempting to see beyond that something–his specific intention.

I believe that every Master lives something unique built into their work, and there is always a way to understand and appreciate it.

When it comes to paintings and visual art, I have a personal method. I sit in front of an artwork for a long time, searching for a connection with the artist. I focus on the details and the overall composition, trying to understand the artist's thoughts and emotions while creating the piece.

The most intriguing part of this exercise is when I try to think beyond what is represented in that piece of work. What other messages are built-in? Of course, Leonardo has always been the most exciting artist to deal with. I have always taken immense pleasure in standing in front of his work, and there is no limit to what can be exchanged with him if you have the wish, the time, and the power of observation in his thoughts, masterpieces, sketches and any work he produced.

First and foremost, it is crucial to acknowledge and understand our own limitations before attempting to leverage our perceived strengths.

I lack Leonardo's constant dedication and commitment, which means I do not have 100% of the time devoted to my mission. I do not have Leonardo's silence and conditions to be not distracted. Furthermore, I don't have the skill to transform any desire into a lasting determination to observe, study, research, create, and invent. Additionally, I do not have Leonardo's good eyesight, which has undeniably limited my ability to observe much of the world around me, day after day, detail after detail.

My curiosity blends well with my imagination, creating a positive impact on my creativity. It all began with my studies in music, fine art, architecture and interior design. Over the past four decades, I have continued to build up my skills, focusing on photography and the ever-evolving digital technology, utilising various computers, applications and peripherals. I have also learned to produce artwork using sophisticated equipment such as laser cutting and water jet machines. Astronomy has always been my passion, and the mysteries of the universe have never failed to fascinate me.

When I envision Leonardo's time, what other factors do I have that support my capacity to create, innovate, and move forward? I have 500 years and counting of more human development, which we call evolution. I have the entire world as a physical dimension at my fingertips, with fast communication, the internet, computers and the entire digital dimension. I am part of a global population experiencing constant logarithmic acceleration and engaging in a globalised competition arena.

I invite you to visit the website accompanying this book by scanning the QR codes found throughout the pages. There, you can explore the various concepts that I envisage and discuss. The content on the website is the result of years of observation, experiments, and experience gained through the development of different projects. It reflects my passion for innovation and my constant drive to challenge existing formulas.

Nothing Is Impossible!

We have no excuses! Just looking at the sheer volume of Leonardo's achievements in so many different fields, with none of the tools and technologies we have today, we could do much better!

THE ART OF PORTRAITS

Let's continue this amazing relationship between who we are and who we see through our eyes, minds and emotions. Leonardo was an absolute genius at representing people up to reaching the absolute enigmatic expression that can mean what you like to see and feel. We can now expand the way we tell a story about someone else.

Art is never finished, only abandoned.
Leonardo da Vinci

A blend of real and hidden mysteries, perhaps even beyond our time.

A portable snapshot of 'my essence', a whole story in a few images.

Rossella: A human chair that carries the essence of a person.

It is a sculpture where you can sit, but preferably better to look at.

TIME MACHINE

We are all like time machines, carrying the stories of our lives and the history within our DNA. We should be able to share our stories to others now and in the future. Technology can significantly enhance our ability to communicate our life experiences with others at any time.

Men spend the first half of their life ruining their health and the second half seeking to heal.
Leonardo da Vinci

Chapter 9

Change

We often take things for granted these days. We have become early adopters of all the 'must have' new shiny gadgets. We have become fixated on social media platforms and share posts, photos, and videos without realising it. This is causing us to self-select which peer group we associate with, and it is almost a separation by online social groups.

Age Over 50 or Under 50

Over 50, you have strong roots and traditions to draw from. We have traditions, habits and a comfort zone. We tend to resist change and stick with what we know. Of course, there is nothing wrong with that, and many people count this as a strength because it comes with wisdom and life experience.

Under 50s tend to have a more open and inquiring mind. They are the early adopters and the embracer of new technology. They do not carry so much baggage and have grown up in a social environment, communicating more online than face-to-face. They are more open to the possibilities of Artificial Intelligence (AI) and less likely to be worried about its potential to disrupt the status quo. They probably would not understand what the 'status quo' means and probably think of the 1967 British pop group of the same name.

Immerse in a different time, into a painting, into a dream.

Generation Z is more concerned than older generations with academic performance and job prospects. Despite concerns to the contrary, they are better at delaying gratification than their counterparts from the 1960s. Youth subcultures have not disappeared, but they have been quieter. Nostalgia is a theme of youth culture in the 2010s and 2020s. As we all know, nostalgia isn't what it used to be!

So, what does this mean in our quest to learn from Leonardo's legacy? Our current generation has accepted the fact that the pace of change is faster than it has ever been. This is the norm. They don't know any differently. They are not aware that the internet is a mere 35 years old. For them, it has always been a part of their lives. We accept each new app and platform as if it were expected and a natural progression of what has gone before.

For Gen Z, AI is just a continuation of the progress they have witnessed. It is just a new tool, a new platform, and nothing to be scared of. Leonardo would have seen it in the same way, as something to be mastered and used.

Over the centuries, changes in technology have affected the entire population at one time or another and have caused entire industries to collapse. The 12th-century landscape was filled with windmills that ground the corn from the fields and turned it into flour, which went to local bakers. Slowly but surely, steam power gradually replaced wind, and the days of the windmills were over. Millers found themselves out of work and looked for new ways of making money.

Travelling between time and dimensions

New industries sprung up, with steam and industrialisation becoming the next big business and major employers. Cotton and the mills were another industry that grew out of change. Then came the oil and electricity industries. It is this constant change and reinvention that drove everything, and it is the same today.

There is no doubt that AI is going to change many industries and replace many jobs over the next few years. Keep your eyes open for trends!

The 50-plus generation had an expectation that a job was a job for life. Their salary would increase at regular intervals, as would their level of responsibility. Finally, they would retire with a substantial pension. Not any longer.

In a far more volatile world, the possibilities of a job for life are very slim. Once again, the pace of change is causing this. You cannot plan for the future of an industry without being able to predict it. No industry is insulated from change, so how can we plan our future? There is only one thing that we can predict: how we will react in any circumstances.

If you were planning a new career today, where would you start? What sector would you choose that could keep you working securely for life? The honest answer is that there is not sector insulated from the effects of change. So what is the answer?

A Portfolio Career is one that keeps you in the driving seat. Instead of looking at a particular job, you focus on a job and an opportunity to learn a specific skill that you will need in your future career, whatever that might turn out to be. As you grow and develop, you will build up a series of skills and experiences that, in the future would make you the best-qualified person for a particular job. Focus on you, your talent, your skill, your attitude, your enthusiasm, and in a crowded job market, your foresight will pay off!

Don't forget that the job you have today may not exist tomorrow. There is no point in associating yourself too strongly with it. Focus on being the best candidate for your next job, whatever that may be.

Also, look out for the Peter Principle. This is about being promoted to the level of your incompetence. For example, you may be perfect in your current role as an engineer. As a result, you may get a promotion because you are good in this role. Then, you get further promoted to a managerial role—with which you have no experience. A good engineer does not necessarily make a good manager. You have been promoted to your level of incompetence! Now, you are no longer as safe in your role as you were when playing to your strengths.

Going back to the concept of the portfolio career, focusing on growing your skill sets, you never get to a point where you are bored with your job. You are seen as multi-talented and multi-skilled in the marketplace. Changing jobs on your terms is far better than trying to find a new position when something has gone wrong with the company or in the marketplace. Your experience of working with different companies and in various countries will be attractive qualities for any new employer.

With AI, things are already moving fast and causing surprising and unexpected changes. You need to be prepared when they happen. Always be a seeker!

Look at how manufacturing has changed in recent years. Initially, we had factories full of workers producing goods. Today, the factories are still there, but they are data centres. No humans are required. The factories of today are run by a man and a dog. The man is there to feed the dog, and the dog is there to stop the man from touching the machines. Changed days!

Many people do not find it credible that, after 500-plus years and in the middle of all these changes, there is anything that we could possibly learn from Leonardo. After all, since then, we have been to the moon and back, and we plan to have our planet populated entirely by robots. We have keyhole surgery, DNA sequencing, and an online digital world. All of this would have amazed him. However, if he were looking at how far we have progressed, what would he see? What have we missed? What basics have we forgotten?

Have we learnt anything from history? Or are we continuing to make the same mistakes over and over again? As each new generation confidently believes it knows better and discards the past for an uncertain future, what have we lost? More than you might think!

None of us exist in a vacuum. We are products of our culture and our generation. For some people, their outlook and way of thinking might be influenced by their life experiences.

They may have lived through the pain of war. Another group might have experienced a financial crisis or poverty. We all have overriding experiences that shape our lives. It was the same for Leonardo.

In his era, the church was one of the biggest influences on everyone, dominating every town. Attending church was still an obligation, not a free choice. The church bells were rung to call the faithful to prayer. The church had the keys to heaven or hell. You couldn't ignore it.

Five hundred years later, we are now in a very different world. The balance of power has changed from being the church to being the world of commerce. Global business and financial institutions are, in effect, the main power in our lives. They have more influence than politicians–even if they think otherwise! Do you believe that Amazon, with its turnover larger than that of many countries and with its massive customer base online, would stand for interference from any country that was not in its commercial interests?

Values have also changed as the world has become more secular. The diversity of cultures has meant that no one faith is dominant. However, as the Christian churches have become less influential, the up-and-coming generation is becoming more spiritual but less religious. Sadly, the churches are no longer being filled the way they were. It is a very different landscape from Leonardo's time.

Schools now give lip service to teaching religion, and there is no longer the level of moral leadership we grew up with. This vacuum for leadership is being filled, not by politicians but by social media and the very recently introduced digital world, which is amazingly only forty years old.

Our desire for stability is made all the worse by the incredible speed at which AI manifest itself in many new areas. Many websites now boast of AI-driven features. People are now too expensive to create content!

These days, if people want to know what to think, they are most likely to crowdsource their views from influencers on social media. After all, why think when you can get someone else to do it for you? But who influences the influencers?

With the transient nature of news and world events, there are few sources of content we can rely on as we did in the past. The tectonic plates we stand on do not provide the firm foundation we have been used to in our lives.

The speed of information travelling worldwide is now measured in milliseconds rather than in the past when it was in hours and days. We don't even need to look on our phones for this as our smart watches light up instantly with news flashes. We are fast moving towards information overload. We increasingly lack the one quality that Leonardo had, which enabled him to achieve everything he did–the wonderful gift of silence and the lack of distraction in his life. Just what you need to be able to think and become a genius.

Creativity

Creativity is the ability to see things that do not yet exist. It is a quality that Leonardo was well known for. His creative abilities covered many different spheres, and he excelled in each one. How can we emulate his example?

When a creative idea strikes you, have you ever asked yourself, 'Where did that come from?' It is as if the idea had just landed from the sky. In fact, its origin is much closer. Our brains have an amazing way of 'joining the dots' and linking seemingly random thoughts to make connections and create new ideas.

Our entire nervous system, including the brain, runs by electricity. The nerves conduct current, which drives every part of our body. When we refer to the 'spark of an idea', we are being accurate because that's what it is.

Some ideas come to us spontaneously, with no effort on our part. We observe what is happening in a given situation, triggering a spontaneous idea. Actually, this probably happens multiple times a day, but usually, we manage to shake it off and ignore it–unlike Leonardo! With so much inspiration coming our way, why are we not rich? Simply because we do not recognise the idea of what it is: a unique and spontaneous thought. Every invention and the largest of projects started this way, with one thought recognised for its potential and nurtured. It would have withered and died in the hot sun if it weren't seen as something with potential. Who knows what that tiny idea might have gone on to become? It could have been worth millions!

Each tiny thought we have needs to be evaluated and tested. Of course, it might come to nothing, but you never know...
To become known as creative, you need to be alert to random thoughts and respect them as gifts. Everyone has them, but not everyone consciously seeks them out. It is your choice.

Frustration

Another source of creativity is frustration! It is like the dog sitting in the sun but finds himself sitting on a nail. The nail is annoying, but not enough to get him to move. He is too comfortable. There are many things in our lives that we find annoying but not annoying enough for us to do something about it. Maybe it is not our problem. Maybe it is somebody else's responsibility, or perhaps there is no apparent solution. Life teaches us that the longer you do sit on a nail, the more uncomfortable it becomes. There is a point where frustration alone will cause you to act. There may not be an obvious solution, so you might have to invent one, but you now have the motivation to create one. Frustration can be a great motivation for action.

Observation

Our eyes are the gateway to our minds and a major source of inspiration. There is a big difference between casually looking at something and proper observation. With the first, we just look with our eyes; with the second, we also engage our brain as well.
If we are to be truly creative, we need to give the brain something to work with. It needs to make connections with what it sees in order to come up with original ideas. To be a true creative, you need to live in the present. From observation comes information, from information comes insights, and from insights comes knowledge. Jean Jacques Rousseau's timeless words were, 'Real wisdom is not the knowledge of everything, but the knowledge of which things in life are necessary, which are less necessary, and which are completely unnecessary to know.' Allow your brain to play and explore even the most random thoughts. You never know what nuggets of gold are waiting to be found.

Inspiration

Inspiration is the father of creativity. It is the process of being mentally stimulated to do or feel something, especially to do something creative. The word 'do' is key. Knowledge without action is of no use. As the saying goes, 'To know, and not to do, is not to know'.

Inspiration can hit us anytime without warning and propel us on a new creative tangent. We might hear a chance word online. It could be a lyric from a song or just a snatch of conversation. Anything that catches us unawares and causes us to pause what we were thinking about and entertain something new. Inspiration parts the clouds and lets the sunshine in. Welcome it! Use it!

Today, we live in a more connected world than ever before. It used to be that something happening around the world could take at least a day to travel around the world. Not anymore! Our always-connected phones, tablets, and wearable devices spring to life the moment a news feed is released. We have become instantly aware of news events from all around the world. We have also become aware of disasters and events that we might be in a position to support or contribute to.

We are at our most creative when we want to help someone or support a cause that we believe in. When you look at big fundraising events like the London Marathon, it is incredible to see the creativity that it brings out from its participants and their families: the costumes, the community fundraising efforts that the event inspires, and the huge sense of being a part of something massively creative. When you come across a need in another person, it does bring out the best in you. Let it flow!

Boosting your Creativity

There are many ways to take your brain to the gym and give it a workout. The more we challenge our brains, the more effective and creative they become. Memory training requires mentally turning images, numbers, cards, and words into pictures. The brain loves pictures and making links between them. By contrast, the brain hates sheets of text; it is boring to the brain. If you want to become more creative, practice mind sports!

Over the past forty years since the mind sport was founded, the volume of information that top competitors can recall has grown spectacularly. Compare this with achievements in physical sports, which can just be measured in fractions of a second.

For example, it originally took competitors at least two minutes to memorise the order of one shuffled deck of cards. Currently, they can achieve this in under ten seconds. The incredible thing is that you do not need to have special skills or qualities. Anyone can do this. All it takes is practice and determination. What do you think this skill would do for every aspect of your life?

Brain for Hire

There is nothing better to stimulate creativity than a client standing in front of you holding a substantial bank cheque. It is amazing how other tasks, appointments, and other now 'less important' tasks can be moved out of the way and for this project to take priority. Leonardo was no exception and was always on the lookout for a sponsor with a need and deep pockets.

As a way of stimulating creativity, there is nothing more flattering then somebody who wants you to do something that only you can! It energises your brain, and it gives it a whole new focus. We can be so shallow! We can also be very quick to put everything that was urgent a few moments ago to one side, but now it is less so. We all like novelty and flattery, which can give our creative juices a new sense of purpose.

Take a word of warning from Leonardo. The clients who you put to one side to pick up a shiny penny haven't gone away and still expect you to deliver! Many of Leonardo's projects that he did on commission were pushed to one side far too often to make way for a new and exciting commission. Always deliver on your promises!

Thinking

There have been many milestones in human evolution—points in history when things have changed or evolved and the world embraced a new direction or technology. Mostly, these big changes happened gradually, and we had time to adapt to new ways. However, time for change is a luxury we may no longer have. It is happening too fast. You need to be more aware and more nimble.

The evolution of Artificial Intelligence has been simmering in the background for around seventy years. Outside of the world of technology, it has been below the radar as far as the public has been concerned. In recent years, this suddenly changed. Instead of being one of those 'Isn't that fascinating?' stories you get at the end of the television news, all of a sudden, it morphed into a scary threat. AI can now generate deepfakes, creating credible videos and pictures of people that you cannot tell from the real thing. It produces audio of celebrities and politicians who are not authentic. Some websites allow you to type in text, which is played back using the fake voice of a celebrity.

There is an American football team that enables you to have a conversation with an AI-created star player who can answer questions. Who is real, and who is fake? In medicine, AI has often proved to be superior to human doctors in diagnosing illnesses using medical data. Yet another human skill possibly lost to machines.

What we are witnessing is a rush in every industry to see what AI could do to save them money, use fewer people, and increase their profits. We are seeing new examples of this every day. This process has not happened in the past few years; it is a disruptive global trend that has only just started. Where will we be in one year's or even five year's time? This will be uncomfortable! Jobs that exist now, including yours, may disappear. You need to look at trends and extrapolate the direction in which they are going.

Companies like Brain Reserve in the USA, founded by the futurologist Faith Popcorn, are now coming into their own. Faith Popcorn was the first to apply the study of Trends to Marketing, making her name synonymous with futurism, creative ideation and business growth. For over 40 years, Faith has led the way in futurist marketing. To this day, her ideas capture the attention of the global press, and she continues to appear on lists of leading futurists. Her first website was a single page; all it said was, 'If you knew what would happen tomorrow, what would you be doing differently today?' This is a very powerful question that is more relevant now than ever before!

Today, she has teams working in every sector, looking for trends and then 'joining the dots' to see where they are going in the future. Long before it became a reality, she predicted that, in an often dangerous world, we would look towards staying home for our entertainment rather than going out. She called this Trend Cocooning. It was featured in her book 'Clicking' (as in clicking a combination lock when the correct combination is entered). She predicted that we would get more things delivered rather than going out to the shops. She was not wrong! Look at how the whole new food home delivery industry has flourished! Now Amazon delivers everything else, sometimes within a day. A very different world from then!

We no longer have the luxury of time to drift through life and wait for new opportunities to emerge. Instead, we need to look at ourselves and see what opportunities will still be available in the months ahead. A year is too far away. Nothing is a given. The future is not within our comfort zone. The past will never return. Now is the time to use our intelligence and creativity to reinvent our future. The most productive thing to do is to draw some of Leonardo's analytical skills and apply them to yourself, your career, your relationships, and your life. Our current profession does not define who we are.

In the UK in the 1960s, there was a television series called 'The Good Life' starring Richard Briers, Felicity Kendal and Penelope Keith. It was about a couple who decided to give up a nine-to-five job in the City of London and become self-sufficient by turning the garden of their house in Surbiton into a tiny farm. In doing so, they had to reinvent themselves and learn many new skills completely.

A great source of comedy! In real life, such a massive lifestyle change would take considerable courage and determination.

However, looking around my circle of friends, I see many people making similar career changes. The owner of a web design company switched to become a health worker. A talented graphic artist became a landscape gardener. People realise that a lifetime behind an office desk is no longer an attractive prospect. Could it be that they are realising that the working desk will not be there forever?

So rather than looking at what you have done so far in your life, maybe start to look at where your real passion really lies. I was once asked, 'What would you be doing now if you knew you couldn't fail?' That's a good question! What is holding you back from making that change and going in a new direction?

The worst thing is to sleep, walk to work, and do a job you hate just out of habit. Step back and look at your life from a new perspective. Make yourself future-proof, and don't get caught out. AI is not far behind!

The Empty Nest

The advantage of age is that, with it comes wisdom. History is the best educator but with every new generation comes the mistaken confidence that it knows best. With the constant excitement of new inventions and amazing technology comes the belief that the lessons of history are no longer relevant. History itself proves that this isn't true. As the saying goes, 'Employ a teenager whilst they still know it all'. It will not last! The older we get, the less we realise that we know!

Every new generation believes it is poised at the start of a brave new world, only to be disappointed when utopia is delayed and progress is thwarted, often by fear of change or lack of political will.

Having lived in a world with wars, financial crises, pandemics, and political disasters, I have learnt the wisdom of acceptance and adaptability. I have been fortunate to have lived and worked in several continents and cultures, and I have learned something from each one. When I was in New Zealand some twenty-five years ago, I realised that the up-and-coming generations were not seeking conventional nine-to-five jobs but were more likely to become freelance professionals with the freedom to develop their skills and experiences in different countries and cultures.

This was certainly not the case with previous generations. Instead of living close to family and relations, the new generations are now using social media and apps like Zoom or Skype to stay in touch. The intimate family Sunday lunch, sitting around the table, is now a thing of the past. Families are often scattered across different towns, countries and continents and may well be aligned with other cultures and mindsets. One thing is for certain: the sharing of ideas, problems and family relationships cannot survive at a distance.

At home in Italy, I live close to a beautiful family-run craft bakery. The inviting smell of fresh bread wafts down the street and draws you in, their croissants make for a perfect breakfast. But will they survive in the future when the founder is no longer there? Will the younger members of the family step up to that opportunity? Will they be prepared to put in the hours and the sacrifice of their parents to build the business in the future? Perhaps not, also thanks to AI.

We will end up with what I experienced in some big cities. You look down the street at the brightly coloured cafes with authentic-looking tables and chairs. But as you look closer you see that they are no longer family-run. The croissants may look the same, but they are now mass-produced. The brand of coffee is the same no matter which cafe you visit. It is almost a Disneyfication of our world. We have been fooled. Many people are no longer prepared to follow in the footsteps of a family business. We are fast living in a theme park as big business takes over.

In Palma de Mallorca, there is the Pueblo Español Mallorca, a Spanish show village created for tourists. It is authentic in every detail, with reproductions of famous buildings from Cordoba, Toledo, and Madrid gathered together with typical houses from the Spanish regions. You can eat Spanish food in the Plaza Mayor or sit outside a cafe watching the tourists buy pearls and souvenirs at the village shops. There are many theme parks like this in the world, and more are coming. Is this the way we are going? Are we moving into a well-presented set-up, where the entertainment is ensured 100%, some level of education is provided, business is optimized, and technology is always implemented at its cutting-edge capacity, with virtual, augmented and mixed reality included?

Living away from home, we yearn for family and the authenticity of life there. Especially the cooking! We desire our homes and take our culture with us wherever we live. We need to anchor ourselves with our original roots. Without an anchor, we are drifting and on our own.

Because we are becoming nomads and no longer live as our parents did, we no longer need a house full of clutter and weighed down with a mortgage. Now we travel light with a rucksack, not a suitcase. We can study and work wherever we like and pick and mix the culture and the friends we choose. We are no longer chained down but free. However, the price we pay is that our morning croissants now come from the wholesalers, not the family baker. It is a sign of the times that in Brick Lane in the East End of London, once a thriving Jewish and then a Bangladeshi community, the very last bagel bakery has just closed. The past is passing as we look away. It will probably be replaced by a branch of 'Bagels R Us'!

Wisdom

Wisdom is a word which seems to have fallen out of favour. Rather than seeking out the knowledge and philosophy of intelligent and experienced people, we look for pithy sound bites and quotes on Instagram and other social media platforms. It is the equivalent of preferring fast food for thought to eat at a Michelin-star restaurant. A snack will give you momentary pleasure, whereas a proper meal will fully satisfy your appetite. More than ever, in such an increasingly fast-changing world, do we need the wisdom of sage minds?

It is a sad fact that when somebody quotes the ancient Greek poet, Homer, most people think they are referring to the television series The Simpsons. But the original Homer said, 'There is nothing new under the Sun'. That piece of wisdom has stayed true over many centuries. Of course, things have changed in terms of technology, inventions, and politics, but fundamentally, man's reaction to changing circumstances has followed a predictable loop throughout that time.

As with the suspended balls of Newton's Cradle, predictably passing their energy from dropping the first ball to the last one jumping, it is very clear that we have learnt nothing from history but seem intent on repeating it time and time again. By doing so, we demonstrate that we would rather ignore the wisdom of the past and look to future 'shiny pennies' on social media instead. As a result, we stopped looking back and, as if wearing horse blinkers, only looked forward. But in doing so, we are doing ourselves a great disservice.

In reflecting on Leonardo, we see somebody who spent his life contributing to the sum of the world's knowledge across so many different areas of science. It was his diligent research that sparked new areas of science to be formed. His voice, knowledge, and experience are as relevant now as they have been over the 500 years since his death.

Leonardo passed on his knowledge and wisdom in his written works, such as his thousands of pages of notes and diagrams, but also in his spoken words. He had no idle academic but was a man driven on a mission, as he said, 'I have been impressed with the urgency of doing. Knowing is not enough; we must apply. Being willing is not enough; we must do.' For him, 'the noblest pleasure is the joy of understanding', and this curiosity was what drove him.

He believed, 'There are three classes of people: those who see, those who see when they are shown, those who do not see.' In his case, his ability to see details was amazing.

He said, 'Nature is the source of all true knowledge. She has her own logic, her own laws; she has no effect without cause nor invention without necessity.'

To him, 'The human foot is a masterpiece of engineering and a work of art.' He proved this in his sculptures, sketches and paintings.

Although Leonardo planned to write a book on art and one on anatomy, his contribution to a published book was his research and illustrations for the work on mathematics, The Divine Proportion, by the Italian mathematician Luca Pacioli. This publication became famous among mathematicians and artists, representing divinely inspired simplicity and orderliness.

Near the end of his life, he was quoted as saying, 'I have offended God and mankind because my work didn't reach the quality it should have.' It is easy for many of us who could never aspire to even a fraction of his genius to write him off as being so great that we could never hope to emulate him. But if we drill down to his wisdom, that is a level playing field.

Wisdom is simply the repeated skill of drawing on your knowledge, experiences, history, and all the life lessons you have learned and using all of these as a filter to pass through your current circumstances and need for answers. You do not need to be a genius to do that, but you do need to do it. You won't get answers if you do not ask the questions. That is the wisdom you can learn from Leonardo!

The Future Beyond Imagination

Every generation has a vision for what the future might look like. From the sci-fi comics that started in the 1950s, we have been filling our fantasies with thoughts of spaceships and little green men from

Mars landing on Earth. Seventy years later, the flying cars that were featured in the comic strips, have yet to materialise. This is a future that was imagined but never materialised, or not yet.

In the Eagle comic of the 1950s, Dan Dare was the 'Pilot of the Future' with his chiselled jaw, peaked cap, smart uniform and sheer determination all over his face. A bust of him can still be seen in Stockport.

His story-lines of the future were entertaining but nothing more than adventures and escapism. The view of the future was based on what we knew and could imagine at that time. As we had no idea of the difference technology was going to make to our world, our ideas of the future seem a little tame today. We did not even have words then to describe the internet or what digital technology has now made possible. Who could have predicted that every home would be digitally connected with streamed video or that we could see and talk to anyone worldwide?

In his cautionary tale, 'A Christmas Carol', Charles Dickens told the story of a miserly businessman, Ebenezer Scrooge, to open his mind to what the future and the present might bring. To do this, he brought him three ghostly figures representing the Past, the Present, and the Future to show him what the future might look like if he hadn't changed his ways and attitudes. How frightening must it have been to him to see that he was the cause of his future misery.

We don't have such a scary vision to jolt us into changing course, or do we? Every choice we make in our everyday lives takes us one step closer to our ideal future or one step further away. Each step is another choice.

Although our future is very much influenced by technology, that is not the only influence on us. Our future is governed by the person we have allowed ourselves to become. As you look around the world, country by country, you do not have to look far to see where idealism and division have the loudest voice, where personal values and freedom of expression are suppressed or ignored, where power is everything, even if it means riding roughshod over democracy. Which voices we choose to align ourselves to will also direct our future. For bad things to happen, all it takes is for good people to turn away and remain silent. Less visible is the hidden power that is holding back the future.

Many inventions that will affect the future are already in place. They have been patented and are ready to bring forward. However, this will happen when the companies involved in controlling these inventions meet their financial profit expectations. One of the drawbacks of this entire evolution and the progress into the future is that most, if not all, inventions are based on precise, accurate business plans and models. This structured formula limits the capacity to open new ways. To be productive and creative, like Leonardo, we all need freedom of thought not formulas that restrict our imagination.

We can only now look back at the Covid-19 pandemic and see the many subtle changes that have happened as a result. We were forced to adapt to staying home and isolating ourselves from the outside world. This broke many habits and forced us to find new communication and work methods. We managed. We changed. And when restrictions were lifted, we decided not to return to our old ways. We chose a different future. We are only now discovering the consequences of those choices. You can never put the bubbles back into a bottle of champagne!

That trend of working from home has continued, albeit for just a few days a week. As a result, companies no longer need to have the same amount of office space. This has affected real estate, fewer people commuted–and effect on public transport. Fewer people ate out at lunchtime, hitting the hospitality industry. Even the smallest change can have unexpected effects. Before Covid-19, could anyone have predicted how it would change these futures?

Fear

People get frightened by many different things. If you were to ask Google for a list of phobias, you would be amazed at how many there are. A phobia is an anxiety disorder involving excessive and persistent fear of a situation or an object. Exposure to the source of the fear triggers an immediate anxiety response. It is a form of mental illness and is far more common than you might think.

All of us have fears, some greater than others. These are not helped by the fact that we live in an increasingly negative world where uncertainty surrounds us. It sometimes seems that there is a conspiracy between the press, radio and television to feed us a diet of doom and gloom. It is as if they have a policy to highlight any bad news and ignore anything else that doesn't match this agenda. Not only that, but social media seems to be out there to shock us, depress us and help feed a climate of fear. Right now, fear seems to be at the top of the agenda. We all receive a daily dose of it.

We might convince ourselves that we are living in the worst of times, but look back in history; there have been vastly worse times in living memories that are easy to forget. It is not the events themselves that are the problem; our perception of them causes us to fear. However, the biggest fear holding people back is a new one. Of course, people are still afraid of heights, they are still afraid of the dark, and the bogyman under the bed! They are still afraid of loud noises and public speaking, but there is now a very modern fear that causes people to stop what they are doing and address that fear multiple times a day. What is it? The Fear of missing out. FOMO.

FOMO causes us to habitually reach for our smartphone to check if something has happened or if somebody has said something. With social media, this has become an epidemic! People disengage from their work, studies and occupations every few minutes–just in case! People now spend many hours of the day on social media because they are afraid of missing out. Platforms like Instagram are very clever about lining up a sequence of captivating video clips to grab our attention. It is very difficult to stop watching!

Leonardo didn't suffer from FOMO; instead, he was at the cutting edge of inventions and innovations that others feared they would miss out on! In Leonardo's time, deadly plagues, battles, wars, and natural catastrophes were equally devastating. But Leonardo looked at things differently; he looked for opportunities in adversity.

When there were wars, he became an inventor of war machines and weapons. He saw a problem and came up with solutions that his sponsors were happy to invest in. He was the creator of things others were afraid of missing out on. He created the FOMO epidemic on his own!

The future is creeping up on us and quietly drawing us into the worlds of our imagination and fantasy. For those who can afford it, immersive worlds will entice us by promising solutions to our deepest desires with no consequences. We will be flooded with choices, experiences, relationships, and entertainment in every form—an online fairground that never closes.

The physical world as we know it today could never compete with such an intelligent and sophisticated world that knows all our secrets, desires, and preferences. It is populated by the perfect people we have created to guide us on our journey and help us find what we are looking for. Dobbie, the house elf in Harry Potter, is a perfect example. You would believe he was real, and yet he only exists on a motherboard.

It will be far more than just having a virtual assistant constantly in our EarPods to talk and look out for us at every moment. With Artificial Intelligence, they are becoming more proactive and able to anticipate our needs. They have our locations and follow our every step. Will they be able to resist telling us about a store we just happen to be passing?

Anyway, EarPods will soon be replaced by a simple implant that ensures we are always connected, not just with audio, but in fully immersive 3D. It just takes a thought to enable us to switch worlds, from real to virtual or a combination of both. Nobody will know if we are talking to ourselves or Siri.

We can go from our present reality into one of several equally credible virtual worlds where we would be known, welcomed, and guided through the shopping mall of opportunities. Our membership opens every door, and our purchases are paid for on our very real account. These worlds are exclusive and only for members who have set up a subscription–closed to everyone else. I can visualise American Express opening an aspirational virtual shopping experience complete with virtual shopping assistants. That'll do nicely!

You can simply navigate between virtual worlds. All your favourite brands will be there, inviting you to explore with a free 'try before you buy' pass. Google will have its portal. It will attract your attention by offering you exactly what it already knows you are looking for. There is no guesswork here.

The old search engines that we are familiar with are long gone. Everything is done by thought or voice. Keyboards and mouses have vanished. Just say the word, and it will appear on the screen. We choose our constant virtual companions with great care. They are in our heads and thoughts 24/7. They facilitate everything happening and indeed, anticipate what we want. We will choose their personality and physical appearance with the same care as we do a wife or a husband. And we won't need to get them drunk first and take them out for a meal!

We step seamlessly from our chosen alternative world into our current physical reality. Our desire for a cup of coffee, whilst negotiating an alternative reality, causes an instruction to be sent to our real-life coffee machine. In fact, everything we need for our physical well-being is linked through a highly evolved internet of things to produce an interconnected smart living space. It knows our routine and our likes and dislikes. It only lives to serve our needs. Everything we need to survive will be delivered to the door automatically. Home appliances, our heating, lighting, doors, kitchen, entertainment systems and security are all connected and await our verbal commands. We have our own obedient world to serve us and with a perfect attitude. No tips are required–just the regular monthly subscription.

Who will be behind these ever-expanding new worlds? We don't need to look far to see the billionaires with the biggest incentives to create virtual worlds to attract us. The biggest of these is the world of gambling. It is glitzy and brash, has the promise of riches, and is addictive. Who could regulate this in a virtual world? With the offer of a little flutter in every game, who could resist?

Universities will no longer need large and leafy campuses. All education can be delivered straight to the individual. All they need to do is log in wherever they happen to be or go to the nearest immersive experience room. There could be a queue.

The world of pornography and the fantasies it provides is already a global industry reaching into the heart and the privacy of our homes. The door to this world is already wide open online and will open even further. It will only become more enticing with high definition and 3D.

Our love of music, live events, stars and celebrities is also already a world we know well. Every show and every performance is already available online. All the platforms will compete to make the viewing and listening experience more enjoyable. Once again, all the choices we make daily feed into the system and create personalised offers and viewing choices. We are being very expertly targeted to influence what viewing products we purchase and consume.

Our constant virtual companion will also be aware of what we like and search for it in advance. She/he will take us on a tour of every virtual shop in every virtual mall. She/he will help us to go virtual sightseeing and suggest places to visit. The virtual experiences created by the major tourist attractions are now better than visiting them in person. Of course, there will be an admission charge to enjoy them but a small price to pay for the experience. Now, there will be no geographical boundaries. We will be able to explore the entire globe online.

By now, we are living in much smaller spaces. We no longer need so much space aside from the basic facilities to wash, sleep and eat. As all food will be pre-prepared, we will no longer need a kitchen. Our choices will be to Eat out, Take out, or Thaw out. Life will be simpler! Restaurants, themed eating places, pubs, and live entertainment venues will attract even more visitors, as we will spend more time in our virtual worlds, we will treasure spending time with real people.

The most important space in our 'living pod' will be where we interface with our lives, which, for some, could be a virtual headset. For others, it will be a video wall or a dedicated digital multi-function room. We don't need a camera because our avatar will display the best version of ourselves. It will always show us as perfect and we will never need a haircut or fresh make-up.

In this space, we can sit and receive our education. The best professors in the world, or their avatars, will talk to us. There is no need to travel to a classroom. Artificial Intelligence will check your essays and evaluate your work! You will be able to download your certificate of qualification. The retinal recognition scan system will certify its genuineness.

There are already many exercise machines that are connected to online virtual trainers. These will get cheaper as AI replaces human trainers with credible avatars. Real trainers will become more expensive as a one-to-one experience will be a luxury service.

Walking platforms mean we will be able to exercise by going for a walk in any virtual park or green space in the world. You will even have a USP fragrance generated to enable you to smell the grass or the coffee!

Some dedicated apps already allow you to play virtual golf, racing cars, fishing competitions, sailing regattas, and many other games with people from anywhere in the world using the same platform–a perfect bunker-free experience on your headset. This environment will keep expanding exponentially.

Forget standing for five hours in Hyde Park watching a concert in the rain; you can now attend a huge concert and enjoy a fully immersive experience without stepping outside. Don't worry, the beer is behind you in the fridge. Alternatively, we can hold a virtual party and meet up with friends from anywhere in the world without having to clean the house afterwards. Don't forget, there is no dress code for avatars!

We will have a full and satisfying life on our own terms without exposing ourselves to the dangers of the outside world. You could be located anywhere as long as you can be connected to the grid. This is true freedom across borders.

New worlds will be created where our good-looking avatars meet socially with others. It's probably best not to meet face-to-face to discover what we really look like–reality can be a step too far!

More importantly, the two biggest challenges we all have as we get older are loneliness and a loss of purpose. Both of these can be cured in our future world. New friends are a click away–and so are virtual AI friends who can chat, and have an infinite capacity to listen. They will not mind if you tell the same story twice. In fact, they will probably remind you if you forget something!

Are you looking for a new purpose? Think of everything you have learned during your life and all the experiences you have had. You have some knowledge that AI needs to be able to fill in the many gaps in its knowledge and to be able to repackage it to help other users searching for that knowledge. AI might even pay you to provide it!

Of course, as with any new breakthrough, it can take at least some years to become mainstream, but looking at the trends can help you to see new opportunities. As the owner of any business you have to ask yourself, what platforms does my business need to be seen on for people to do business with us, in the future? A website builder will not be enough and is now yesterday's technology. In your sector, what existing services or experiences will probably be delivered online in the future, probably using AI experts or presenters? Look beyond the present. What direction are things going? Where are the new opportunities?

Most of what we have described is already in existence. It has been developed and it works. All that is lacking is the belief that this is the future. One thing is for certain: the billion-dollar businesses that shape the future are working in this space to create the interfaces that we will need to join. If only, in the future, disputes between countries could be settled by a virtual war where one platform competes with another where the prize is membership, without anyone getting hurt. But in the future, anything is possible.

The War of the Worlds

The most interesting battles in the near future are not going to be with warring factions, dictators or ideologies. Instead, it will be with ourselves and how willingly we embrace the merging of our physical and virtual worlds. It will be in the ease of switching from one to the other and our willingness to trust the virtual space. The physical and virtual coming together has already partially happened, but the full development and implementation are being held up because the potential of the combining worlds has not yet been understood fully. The future is tapping its fingers, waiting for us to catch and embrace what is waiting around the corner.

The mirroring interconnected real and virtual worlds environment is also being held up because the mega companies that have the resources to make it work probably have their vision and financial interest elsewhere. Let's hope we won't need to wait much longer. These new virtual worlds will be instantly addictive, as the current virtual gaming world has already been proven and safely hidden behind passwords and biometrics. No gatecrashers here. The landscapes will become ever more lifelike, and will the people, or rather, avatars that we will be conversing with. As we navigate our way around the virtual town centres, entertainment centres and retail parks, we will all seem strangely familiar. This is because, with Artificial Intelligence at its heart, it already knows all of our preferences: Where we like to visit, what we enjoy doing and who we like to spend time with. This world will be so inviting that, every morning, as we wake up, we can't wait to log in and pick up from where we left off.

As we walk past a virtual art shop (which is, of course, no accident!), we will be greeted by a very realistic virtual expert outside the shop who will engage us in conversation and invite us in for a guided tour. We can inspect all his artworks in detail and even make a purchase. This is where the virtual and real world join together to link up with our real-world payment platform and fulfilment service; the art piece will be dispatched to our front door. You will be able to call up your friends and invite them to join you in the virtual shop to get their opinions on your potential purchases. Certainly, the quality of avatars has dramatically improved and has become so lifelike. There are already companies able to scan your real body and all your facial expressions and deliver to you an avatar that mirrors the real you.

In fact, as we immerse ourselves in these virtual worlds, our experiences will become more lifelike every day. We will be able to attend virtual concerts and festivals without the famous mud from Glastonbury. We will be able to have a virtual seat at any number of conferences and training sessions. Probably, the speakers will be AI-generated.

Will the schools of the future also be virtual? Will we log on to our virtual GP? The future has infinite possibilities. But stop! Are we focusing on the wrong thing? It is fun to speculate on the future, but we all must live here and now. We need to earn an income. We need to live irrespective of what technology is doing. If we have learned anything from Leonardo, it is that he very much lived in the real world. He was consumed with curiosity and was constantly learning. He was fascinated with nature and art. He looked for problems to solve. As an entrepreneur, he was always looking for clients and sponsors.

Would he have achieved what he did if there had been social media five hundred years ago? Like so many of us, would he fritter away hours engaging in insignificant conversations with people he has never met, or scrolling through time-wasting videos of grumpy cats? If he had done, would we today be able to admire the Mona Lisa? Would he have had time to paint the Last Supper? Might he have never painted Salvator Mundi or the Virgin of the Rock because he was busy on Facebook?

It sounds ridiculous, I know. But if we were to look at our own lives, and the actual time we spend on our smartphones, what else could have we done, been to, or achieved in that length of time? We have forgotten that life is a gift and that tomorrow is never guaranteed. We only have the now—and if we give this to an inanimate device which has no brain and no soul, we are left with a growing list of all the things we should have done, all the people we should have meet, all the things which would be our legacy.

If Leonardo had a smartphone, his name would most likely have been long forgotten and his achievements would probably never have happened. What about us? Are we prepared to waste our lives? What big achievements will we leave without starting? Will our lives be celebrated, or will we just be a cautionary tale told to our grandchildren?

The good news is that we can do something about it now. Look back at these chapters and ask yourself, 'If I had the time, what could I achieve?' Then switch off your phone and discover that you do have the time!

The most difficult thing you could ever do in life is to decide to disconnect yourself from that all-pervasive digital world, even for a specific length of time. All you need is time, an open mind, and a clear purpose. Sadly, being disconnected even for a minute is becoming almost impossible to do. I dare you to try it!
To achieve anything, you need quality time for your brain to do what it is best at without interruption. You will discover that, like Leonardo, there is nothing you cannot achieve if you turn your mind to it!

Grumpy cats will never put food on your table!

Experience is the only teacher whom we can trust.
Leonardo da Vinci

ARCHITECTURE OF THE FUTURE

If virtual and augmented reality become more integrated with the real world and if Artificial Intelligence will optimizes and changes the way we live, how should we evolve in the conceptual thinking and design of the spaces we leave, work, and move into and around?

The human foot is a masterpiece of engineering and a work of art.
Leonardo da Vinci

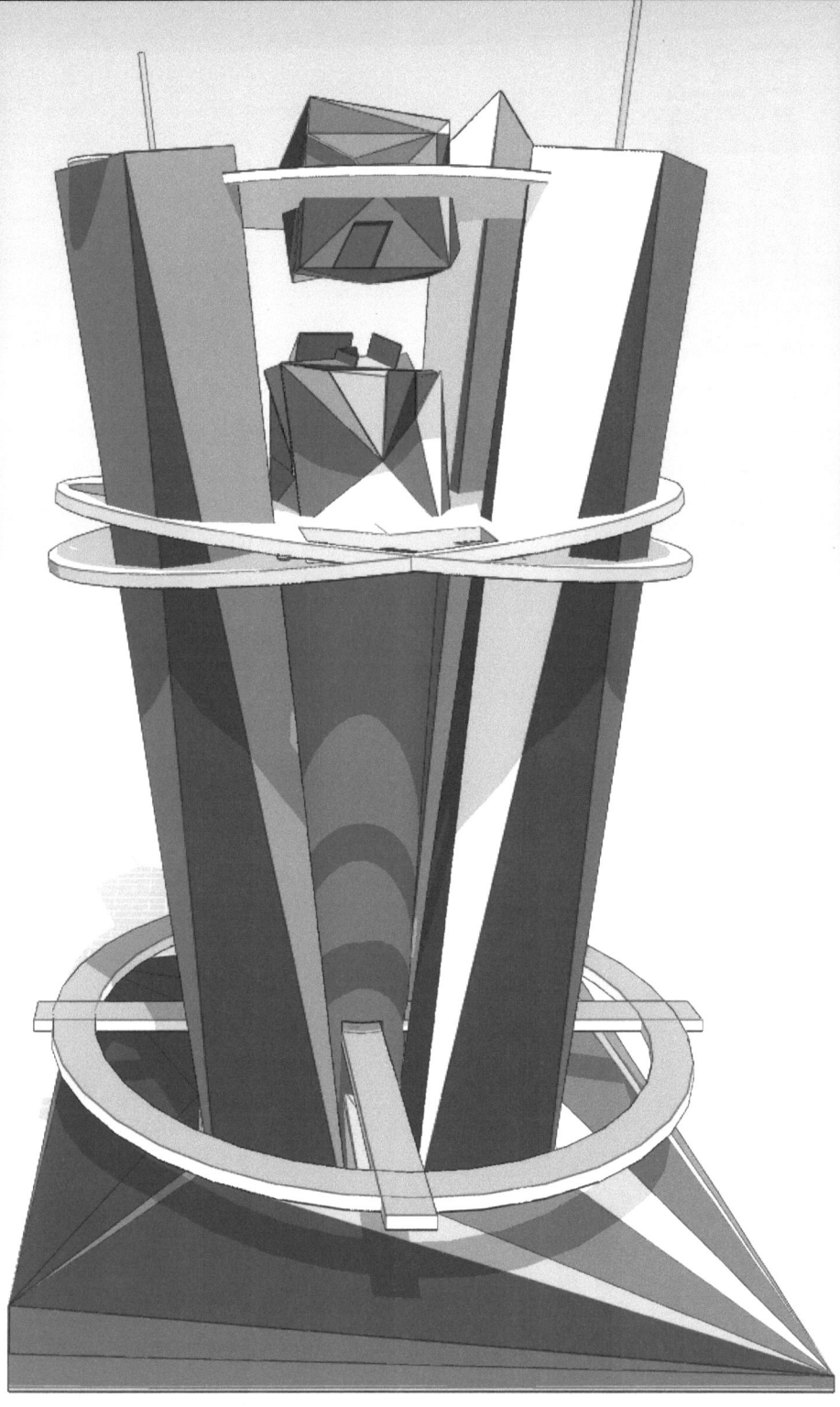

THE CUBE CONCEPT

CONCEPT 7, REAL AND VIRTUAL, ALL IN ONE

Adding dimensions in order to add intelligence and opportunities. The 'game' is becoming more sophisticated and challenging, so now we should push ourselves to use much more of the capacity of our brain.

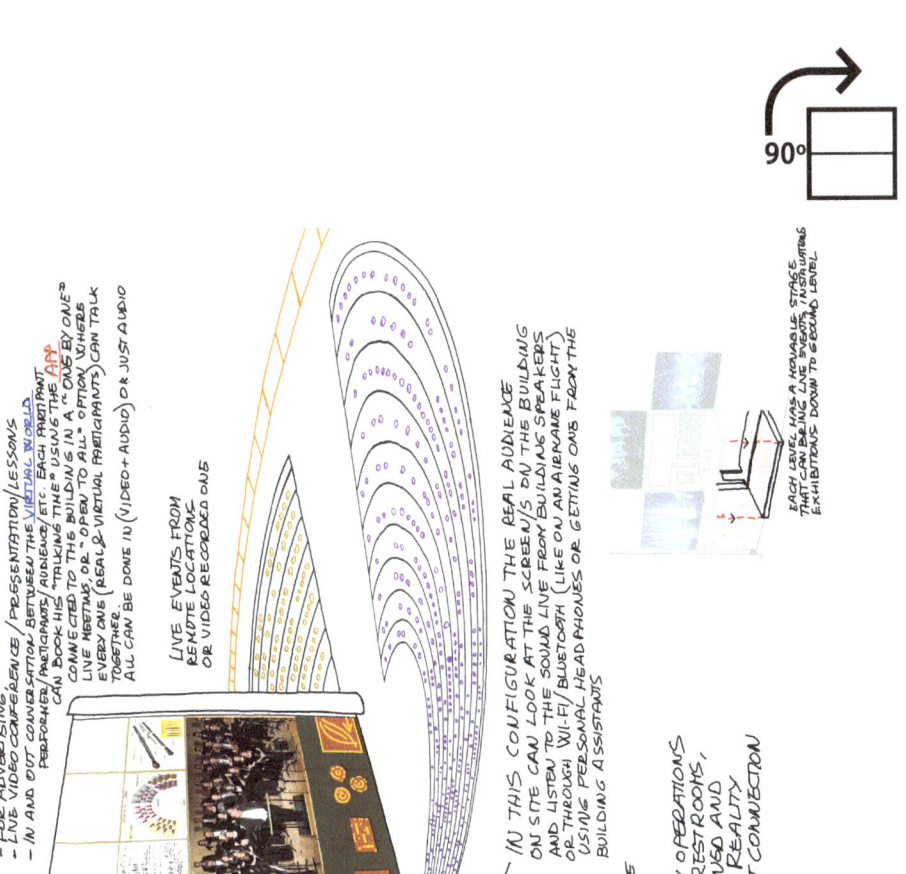

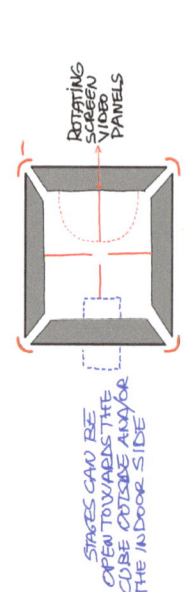
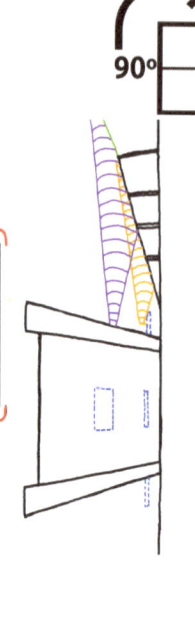

THE CUBE

GENERAL OVERVIEW AND INDOOR FUNCTIONALITY

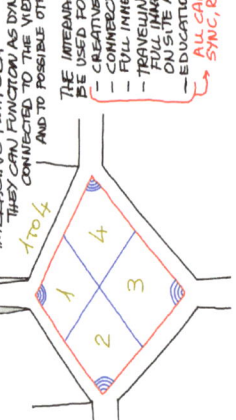

INTERACTIVE PLATFORM
THEY CAN FUNCTION AS DYNAMIC MEDIA PLAYER CONNECTED TO THE VIRTUAL MIRRORED "CUBE/S" AND TO POSSIBLE OTHER REAL "CUBES"

THE INTERNAL VOLUME OF A "CUBE" CAN BE USED FOR
- CREATIVE LABORATORY/WORKSHOP
- COMMERCIAL EXPO
- FULL IMMERSIVE AR & VR SPACE
- TRAVELLING AROUND THE WORLD
- FULL IMMERSIVE LIVE WITH TOURIST GUIDES ON SITE ALL OVER THE PLANET AND BEYOND
- EDUCATIONAL AND TRAINING CENTRE

→ ALL CAN BE DONE CONNECTING IN PERFECT SYNC, REAL PEOPLE INSIDE THE REAL CUBES AND "AVATARS" AND OTHER PLAYERS COMING FROM THE "VIRTUAL DIMENSION"

STAGES CAN BE OPEN TOWARDS THE CUBE OUTSIDE AND/OR THE INDOOR SIDE

ROTATING SCREEN VIDEO PANELS

AN INTERNAL SPACE FOR:
– SPECIAL EFFECTS MIXING OLEOGRAMS WITH SOUND EFFECTS, 2D AND 3D PROJECTIONS.
– IMPLEMENTATION OF MULTI LEVEL USE OF AUGMENTED REALITY
– MIXING EFFECTS COMING FROM MIRRORED CONNECTED, LINKED
SIMILAR "CUBES"
– THE MAIN WALLS CAN ALSO SHOW REAL IMAGES COMING FROM LIVE CAMERAS CONNECTED FROM ALL OVER THE WORLD
3DIMENSIONAL AND OLEOGRAM MACHINES
SYNCRONIZED PROJECTORS ALSO TO BE USED FOR AUGMENTE REALITY AND 3 DIMENSIONAL PROJECTIONS
2D AND 3D PROJECTIONS
MOVING PLATFORMS, UP, DOWN, HORIZONTAL, VERTICAL
ROUND MOVING PLATFORM WITH MECHANICAL SPECIAL FUNCTIONS
TIMELINE
REAL PEOPLE MEETING WITH AVATARS AND OTHER TYPE OF VIRTUAL VISITORS
GLASSES/ HELMET WITH GEO POSITIONING+ 3D SPACE LIVE SCANNER AND 3D PROJECTOR 360° SOUND SYSTEN
REAL PERSON LOCATED IN THE REAL CUBE/ REAL WORLD
AN AVATAR OR VIRTUAL VISITOR CAN "ENTER" IN A REAL WORLD USING THE 3D ENVIRONMENT MODEL CREATED LIVE FROM GLASSES/ HELMET WITH "3D LIVE SCAN" EQUIPMENT
THIS AVATAR/VIRTUAL ENTITY WILL BE VISIBLE BECAUSE "PROJECTED" BY THE "GLASSES/HELMET WEARED BY THE REAL PERSON LOCATED IN THE REAL SPACE
– MANY AVATARS/VIRTUAL VISITORS CAN BE VISIBLE AND IN MOTION AT THE SAME TIME INSIDE THE "CUBE"
THE 3D LIVE SPACE SCANNER RE-CREATE IN LIVE MOTION THE 3D SPACE WHERE THE REAL PERSON IS MOVING INTO AND SEND THE 3D MODEL LIVE TO THE "AVATAR LOCATED IN THE VIRTUAL DIMENSION/WORLD.
THE AVATAR/VIRTUAL PERSON WILL BE PROJECTED FROM THE GLASSES/HELMET, IN FRONT OF YOU OR AROUND YOU, AS HE IS MOVING LIKE YOU IN THE SAME SPACE, THE VIRTUAL SPACE, MIRRORING THE REAL WORLD
AVATARS AND OTHER VIRTUAL VISITORS CAN BE PROJECTED BY GLASSES/HELMETS AND ALSO APPEARING ON "DEDICATED WALLS, PANEL, ELEMENTS
THESE WALLS/ PANELS CAN ALSO BE VIRTUALLY PROJECTED INSIDE THE EMPTY SPACE AND MOVING
THE INSIDE VOLUME OF THE CUBE CAN BE CONFIGURETED AS FOR A MULTI DIMENSIONAL PRINTER, CREATING PARTS FROM VIRTUAL TO REAL
THE INSIDE VOLUME OF A CUBE CAN ALSO FUNCTION AS A FULLY MOTORIZE COMPUTER CONTROL "STAGE" WHERE 3D VIRTUAL INSTALLATIONS, CONSTRUCTIONS, ANIMATED IN MOTION GAME SETS, CAN BE BROUGHT TO THE REAL WORLD BY READING A 3D VIRTUAL MODEL, SPACE, THE CUBE COMPUTER SYSTEM WILL BE ABLE TO MOOVE HUNDREDS OF PANELS AND ELEMENTS IN ORDER TO RECREATE THE SELECTED VIRTUAL 3D MODEL. THIS WILL INVOLVE THE USE OF "ARTIFICIAL INTELLIGENCE" ROBOTICS, AUGMENTED REALITY AND APPROPRIATE "SOFTWARE". THE GOAL IS ALSO TO CREATE A HIGH LEVEL OF INTERACTIONS AND CONNECTION BETWEEN THE VIRTUAL AND THE REAL MODEL
INTERACTIVE LIGHTING, USABLE BY REAL AND VIRTUAL VISITORS
ACTIVE CAMERAS VISIBLE AND USABLE INTERACTIVELY BY THE REAL AND VIRTUAL VISITORS
AUGMENTED REALITY MARKERS
REAL STRUCTURE BUILT FOR CREATE A STAGE IN 3D WHERE THE AUDIENCE "REAL AND VIRTUAL CAN MEET AND USE AR
CAN ALSO BE USED FOR "VIRTUAL TOURISM, THE SCENES CAN BE PROJECTED TRIDIMENSIONALLY BASED ON DATA BASE MODELS OR BY LIVE 3D SCAN DATA, SENT BY A LIVE OPERATOR LOCATED ON SITE, LIKE FROM PARIS
IF 'ORANGE' REPRESENT REAL STRUCTURES BUILT ON STAGE THE GREEN ARE VIRTUAL ADD ONE STRUCTURES MATCHING THE REAL STRUCTURE ON SITE AND THEY ARE VISIBLE USING AUGMENTED REALITY GLASSES AND OTHER AR GEARS. IF INSTEAD 'ORANGE' REPRESENT VIRTUAL STRUCTURE VISIBLE WITH AR GLASSES, GREEN IS ALSO REPRESTING MORE EXTENDED TRIDIMENSIONA STRUCTURES, ANIMATED, INTERACTIVE, TRANSFORMABLE
90°

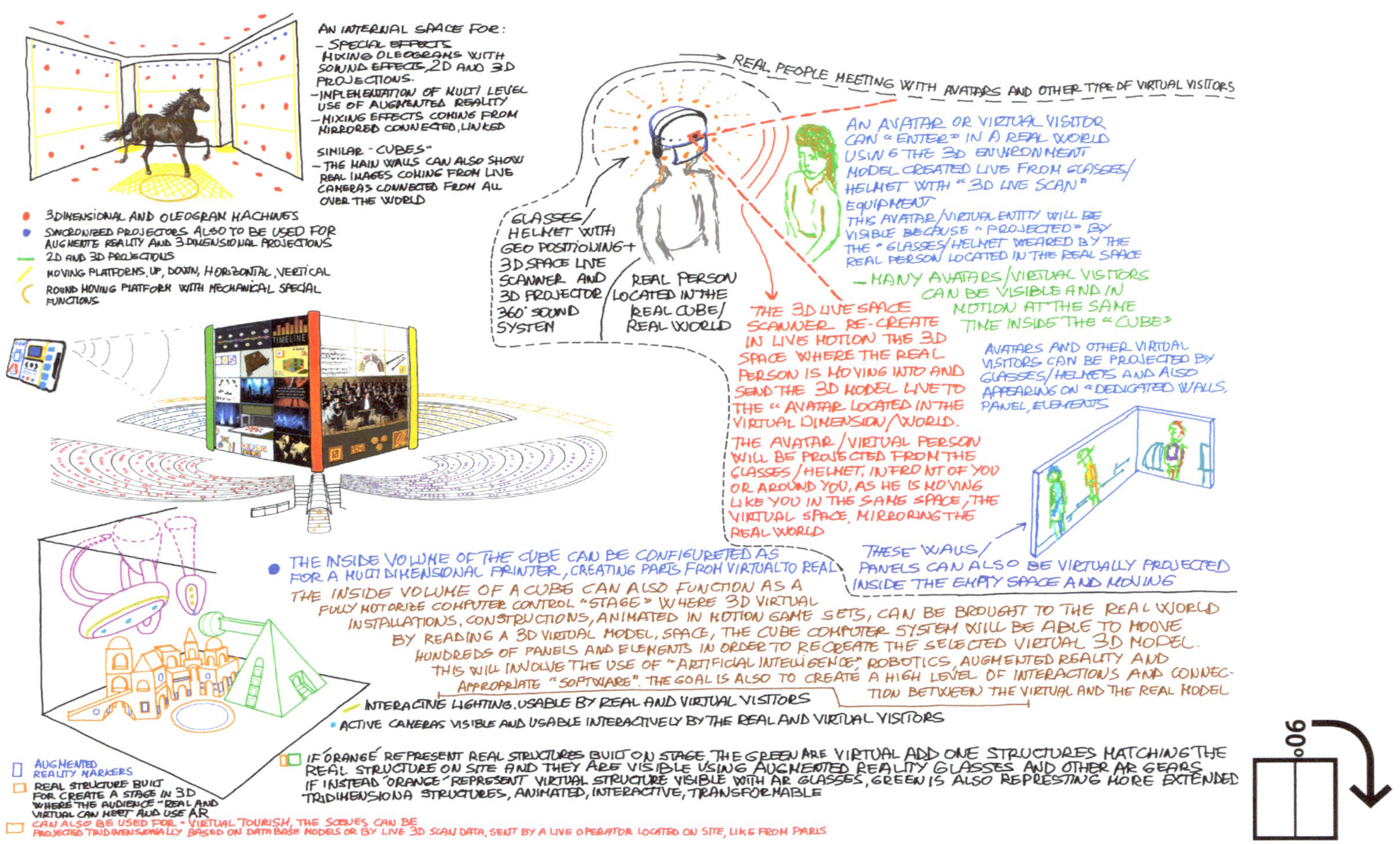

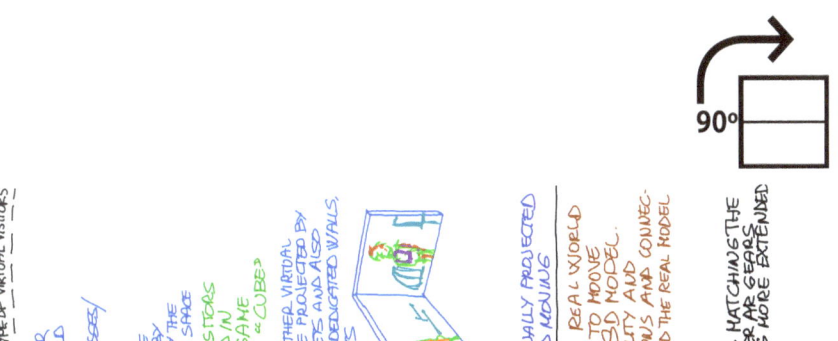

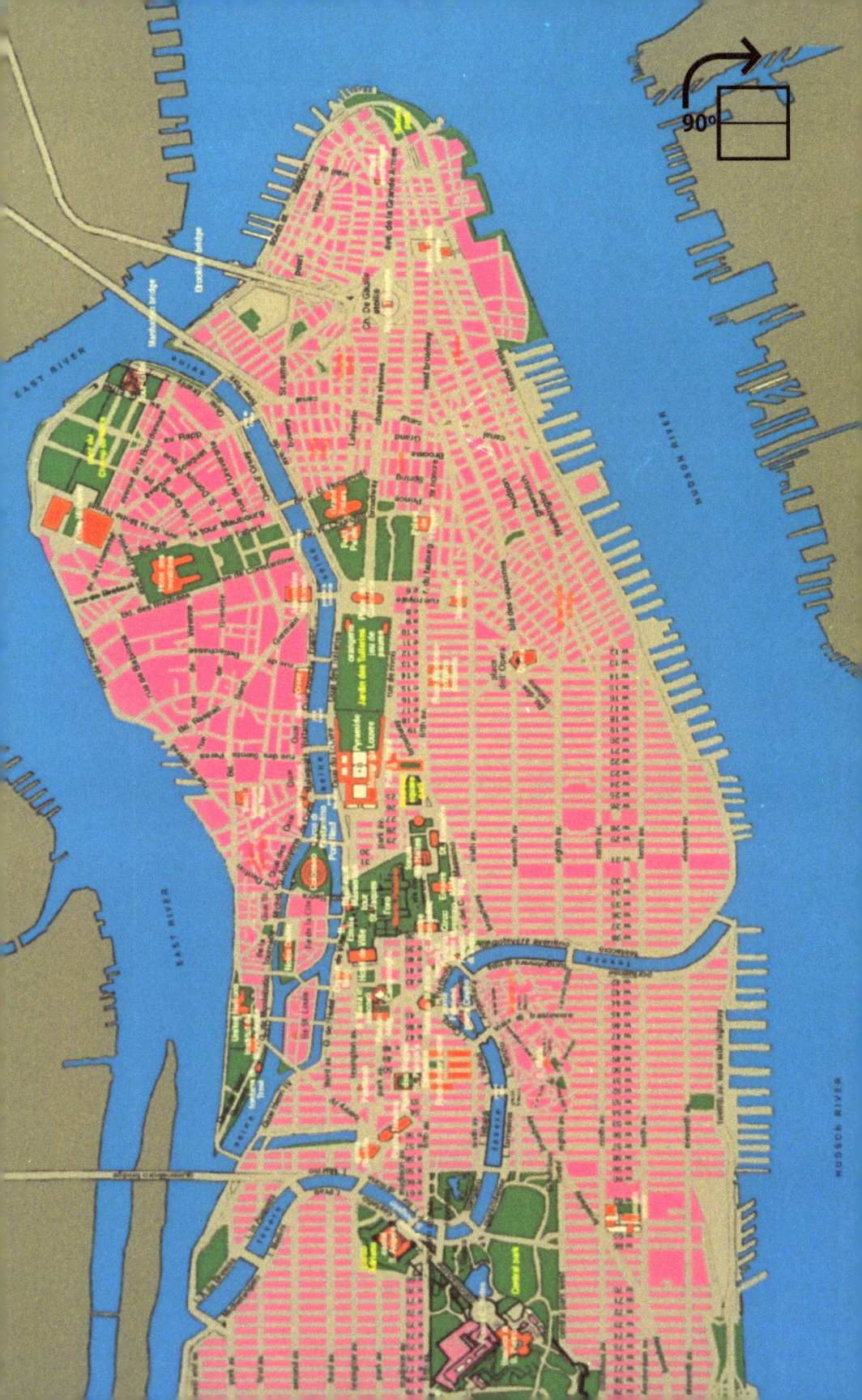

Paris, Rome, and New York are in a real and virtual new arrangement.

All can be combined for us to live in and operate from.

The time machine we all are, 40 years in New York City

Tears come from the heart and not from the brain.
Leonardo da Vinci

APPENDIX

Nostalgia Isn't What It Used To Be!

In this book, we have spent much time looking back into history. We have looked in-depth into Leonardo da Vinci's life, qualities and time. In doing so, we have explored his influence on us in multiple fields of science and art. We have also looked forward into the future to see where Leonardo's influence may take us.

But the most important area we have left until last is the present. What we are doing right now will make the greatest impact in the future—not in theory, not just as an academic exercise, but in the practical way we focus our lives to achieve the best outcome.

We are in the best position to make a difference right now. The past has gone, and the future is increasingly uncertain. If we are to make an impact, the time is now.

Step back for a second and look at the incredible tools and resources we have at our fingertips. We have all the world's knowledge at the click of a mouse. We are connected to the World Wide Web, allowing us to locate and communicate with anyone in any country. Using Google Street View, we can travel around the streets of any town or city worldwide, even the tiny town of Vinci! Try it!

We can look at the lives and works of great scholars, engineers, scientists, architects, authors, artists, statesmen, and great leaders. What a wonderful opportunity! Should we wish to, we can locate and view virtually every past television program or watch a massive archive or sporting event.

These incredible resources are there waiting for us to help us in whatever we choose to do. Leonardo had none of this. He had no electricity apart from anything else! And yet, his numerous creations are still marvelling at us today. He only had his brain and great curiosity. With those alone, he impacted the world like no one had done before or since.

Who will follow in his footsteps today? Will you? Could you?

Strangely, with the many advantages we have today, it is far more difficult for us to separate ourselves from the noise and pace of life so we can have the time to think and be creative. Our lives are now so complicated and interwoven that we no longer have the freedom to follow our thoughts and ideas. We are carrying too much baggage. Leonardo's 'more simple' life, which provided him with the freedom to produce such a mountain of achievements throughout his lifetime, is impossible for us.

Instead, we should apply everything we have learned about his philosophy, to the way he approached every idea and opportunity. If we apply it to just one thing, we will have done well. Realistically, we may not be polymaths or masters of multiple subjects, but we can be the master of one.

Choose wisely what that will be, and then single-mindedly focus all your creativity and energy towards achieving it.

You will encounter intense pressures that can divert you from your goals. The constant pull of social media drains your time, while the Fear of Missing Out distracts you, making you compulsively check your smartphone. Additionally, the allure of live-streamed entertainment adds to this distraction. It often feels like there's a conspiracy designed to steer us away from our chosen purposes. To leave behind the legacy you are capable of, you must overcome these distractions. Will you be remembered in 500 years like Leonardo da Vinci? Only time will tell. The choice is yours.

The Leonardo Mindset

When Sir Christopher Wren built St Paul's Cathedral in 1675, he would have been delighted for somebody to have described it as 'awful'. That is because this word, like so many others, has changed its meaning over the years. Today, the meaning of awful is 'extremely disagreeable or objectionable'. However, in the days of Sir Christopher Wren, 'awful' simply meant 'full of awe', full of inspiration. A thing of great beauty. Indeed, St Paul's Cathedral remains to this day.

So many words have been hijacked and given new meanings by popular culture. It used to be the case that 'wicked' meant 'something very bad'. Perversely, in today's teen talk, it also means the complete opposite.

In these fast-moving days, have we lost the ability to be awe-inspired? Or do we just take everything for granted?

To us, it appears that technology has the answers to all our problems. We are no longer needed. All we need to do is wait at the bus stop for the next great invention to come along and then adopt it without question.

With medicine, give it another couple of months, and gene therapy will solve the problem. Technology? A new app will solve the problem. We are not involved with the process. We have become observers rather than participants. Everything is now somebody else's responsibility. We have become passengers. Far removed from the way that Leonardo looked at the world around him.

Leonardo was never 'an audience', applauding what others were doing. He was the central character in his own drama. He fulfilled his own needs. He took complete responsibility for his world. He did not have Uber to phone so they could deliver life essentials to his door. Today, we are in one giant comfort zone. All our needs are met. We have no incentive to contribute to the world around us. It will supply all our needs. All we need to do is do our job, not make waves, accept when we are given, and keep our heads down. And this is the truth for many millions of people. But not you.

You are different. You have a brain. You are not satisfied with the world around you. It is not perfect. It needs to change. All of your needs are not met. Nobody will ride in from the sunset with a suitcase full of solutions. You will have to find your own suitcase or even build one from scratch, as Leonardo would have done.

The worst thing is that we have lost our sense of curiosity. If we need to know anything, we simply pick up our smartphones and Google it—anything, except work it out ourselves!

Curiosity is the driver of improvement and innovation. It is not something you can delegate. You can't say, 'I am curious about who will sort out this problem.' No! If your brain has made connections, joined the dots, and come up with a solution, with that inspiration also comes the responsibility to see it through to a conclusion.

Leonardo didn't wait for somebody to tell him how to design a helicopter. He just drew on his imagination and started inventing. Of course, you will not solve the world's problems alone, nor will you simply be a detached observer.

If you are to benefit from Leonardo's legacy, keep him in mind whenever you notice something that could be improved or if you have the spark of an idea. Don't just shrug it off and say, 'How fascinating!' And walk away. Instead, look at that moment of inspiration and use it to make a difference. Draw from what Leonardo did and how he applied his philosophy to all the day-to-day opportunities around him. Don't be remembered for all the things you didn't do.

Observation

The Great Illusion—can you really believe your eyes? If we can trust anything, we can always trust what we see, or so we think. The reality is that what our eyes actually see is turned upside down by the optics of the eye. The brain sees everything upside down and then needs to turn it the right way up for it to make any sense to us at all. Don't ask me why! All I know is that our eyes are far more easily fooled than we might imagine.

In 1872, Leland Stanford was looking to settle a bet. The businessman and racehorse owner had wagered that a horse in trot had all four legs off the ground for a brief moment in time, one that passed too quickly for the naked eye to see. Stanford sought out a well-known California-based photographer to see if this relatively new technology could help him win the wager. The resulting work was a shift in Eadweard Muybridge's career as a landscape photographer and became the launching-off point for the first scientific study of motion using photography.

From 1883 to 1886, Muybridge produced over 100,000 images of animals and humans in motion, occasionally capturing what the human eye could not distinguish as separate moments in time. Watching these images in close succession fools the brain into believing it sees the horse or person in motion. It is on this principle that motion pictures were born.

In London, the Museum of the Moving Image curates the development of all the different ways of recording and displaying moving images, from the earliest examples to today's 3D films. As the saying goes, 'You'll believe a man can fly!'

There was once an experiment where a group of dancers were filmed performing. The audience was asked if they spotted anything unusual. They didn't. They then played the film again, and in the background, they saw a man in a monkey suit dancing along in plain sight. Nobody had noticed. Our eyes seem to see what they expect to see. In this case, everyone was focused only on the dancers.

Another factor governing what our eyes see is our subconscious mind. It continuously searches for things we are interested in, which it flags up for us. Our eyes rely on our brains to focus on what we want to see.

The trouble is that today, we do not allow ourselves the luxury of 'looking' in our fast-moving life. All we have time for is a quick glance around us. We are too busy looking around and trying to take in everything in case we miss something. FOMO once again. This is the complete opposite of the way Leonardo used his eyes. He could spend hours looking at nature and transfer his detailed drawings into his notepad. Fortunately, they are still available for us to study today. We can still see the same minute details he saw if only we took the time. But we don't because we are too busy and in a rush.

His paintings are a testimony to his skill in observing. The closer you look, the more detail you see. That is why his paintings are recognised 500 years later as masterpieces. Today, very few have the skill to capture a plant or a landscape on paper or even on a digital canvas. Instead, what we do is take a selfie with our phones. The paper that Leonardo used has stood the test of time, whereas our rushed digital photos will probably not. Devices fail, we forget to back them up, and they get stolen. We lose them. The data is no longer readable. But we still have our eyes and our memory. If only we trained them as thoroughly as Leonardo, we could also ensure that our lives are captured for posterity.

Currently, our capabilities and skills are under threat due to the growth of Artificial Intelligence. Realistic, high-resolution digital images and videos can now be created from the simplest of commands, and they are good enough to fool everyone. Is this march of progress a good thing? Possibly not if you are one of the thousands of illustrators and designers whose career AI might replace. But there is something worse.

As we lose yet another skill to machines, the unique differences and abilities that we all have as human beings are being eroded. Why learn skills that will no longer be needed? We all need purpose in order to thrive and have meaning. Our biggest asset is our eyes and their ability to not just see but interpret the world around us and be inspired to make it a better place.

The fundamental skill we have is our ability to 'join the dots' and to create something new. That ability draws from our personal knowledge and experience, which are unique to each one of us. As long as we have freedom of expression, Artificial Intelligence will just be a tool we can use, not a threat to replace us and our contribution.

It is better a small certainty than a big lie.
Leonardo da Vinci

> *Painting is silent poetry,*
> *and poetry is blind painting.*
> *Leonardo da Vinci*

IN SYNTHESIS

I

Wide as the Oceans
just as changeable
a mighty Kung
representing us in the spheres : sperm whales.

II

(to this purpose we submit
a modest representation
of our sounds;
no more than a note)

III

Mighty. The wait evoked by
wax
on the floors. A tail.
A Kung. Yes, a call.

> *If you can't do what you want,*
> *then do what you can.*
> *Leonardo da Vinci*

MONTGOLFIÈRE

We count atoms of helium.
Morning, it
is grey sanddust on the skin of Beauty
upside-down drops of black
black and white photographs
it is montgolfière :
of rise regulating dosage
and fall from the stars
it runs on thread, plays with flames
stronger, less
zero neuter atmosphere
transparent already of yesterday.

> *Iron rusts from disuse; water loses its purity from stagnation...*
> *even so does inaction sap the vigour of the mind.*
> *Leonardo da Vinci*

IDENTIKIT

Discovering galaxies is my mission
gathering moon dust
I beg for astrospacial crumbs

and sleep on a cushion of notes
attached to a thread
from which I'm nourished effortlessly

Son of a star, I search for the existence
of living souls with the only company
of my Master's rhythmic beat.

Thousands of light years away, a second ago,
there were earth colours and leather boots, meat
leftovers. But I have a mission.

LEONARDOISBACK.COM STORE

The greatest genius Leonardo, Vinci, Tuscany, all of us today, generations to come and humanity. We are all here in our evolution, which will never end until... Let's continue being inspired by the great Maestro Leonardo and keep creating, inventing, and moving forward, astonishing ourselves.

Sad is that disciple who does not surpass the master.
Leonardo da Vinci

An art gallery welcoming real and remotely located artists and visitors.

A Letter From The Future

George Orwell wrote his dystopian novel and cautionary tale '1984' in 1949 which looked about 35 years into the future. The book centres on the consequences of totalitarianism, mass surveillance, and repressive regimentation of people and behaviours within society. It paints a bleak picture. When the year 1984 arrived, people couldn't help but compare his predictions with what actually happened.

In fact, many books or films that have tried to predict the future got it badly wrong. This is simply because it is impossible to predict new technologies that don't currently exist but have the potential to change our lives dramatically. The internet is just one example. Artificial Intelligence is another.

So, instead of looking forward, let's look back thirty-five years to today and see what might have changed. A letter from the future...

It Is All In Our Minds

In the 1970s, there was an American Sci-Fi action adventure service called The Six Million Dollar Man. It told the fictional story of USAF Colonel Steve Austin, portrayed by actor Lee Majors. After being very seriously injured in a NASA test flight crash, leaving him a complete cripple, the decision is taken to rebuild him—at great expense, Six Million Dollars. He was given superhuman strength,

Who does not value life does not deserve it.
Leonardo da Vinci

speed and vision with a series of futuristic bionic implants. This fusion of man and machine was the vision of the future just fifty years ago. But so much has happened in those fifty years to bring this much closer.

Today, this is known as the move towards 'Singularity'–the move towards merging man and machine. It is happening in many ways that promise to enhance human consciousness and cognitive abilities dramatically. The boundaries between what is a human and what is a machine are starting to merge. What ethical questions might we be facing very soon?

As the 2024 Paralympics in Paris demonstrated, sophisticated arrays of prosthetics and artificial limbs were used by athletes to not only replace lost functionalities but dramatically enhance them. The Six Million Dollar Man is already here!

Elon Musk, one of the world's most famous entrepreneurs, announced in July 2024 the unveiling of a new type of Brain-Machine Interface (BMI). His company, Neuralink, implanted a chip in a human for the first time. He is advertising for volunteers with severe mobility and motor skills challenges.

Noland Arbaugh was the first human to receive a Neuralink brain implant chip in January 2024. Arbaugh is a 30-year-old quadriplegic man who became paralysed after a diving accident. The Neuralink implant is a coin-sized device placed under the skull and contains a battery, processing chip, and 64 threads connecting to the brain tissue. The threads have over 1,000 electrodes that can read brain activity and connect to a smartphone or computer.

Arbaugh trained computer programs to translate the firing of his brain's neurons into cursor control.

The emerging market of brain-computer interfaces is in the process of finding its footing. In a world where AI is on the rise, a Brain-Computer Interface (BCI) allows for telepathic control of computers and wireless operation of prosthetics. Musk believes, 'If humans do not enter a symbiosis with artificial intelligence (AI), they are surely be left behind.'

In 2024, news broke of an extremely tiny computer consisting of 8 electrodes. It wasn't connected to a silicon board but to brain cells. Researchers from Swiss biotech startup FinalSpark said this is their new biocomputer, one that mimics the neural network of the brain and could be the future of computing.

So, where did the field of science start? Incredibly, with Leonardo da Vinci. His innovative research on the brain resulted in significant neuroanatomy advancements, including identifying the frontal sinus and meningeal vessels, as well as contributions to neurophysiology. Notably, da Vinci employed a method of injecting hot wax into the brain of an ox, producing a cast of the ventricles. This technique marked the first documented use of a solidifying medium to define the shape and size of an internal body structure.

Leonardo developed a unique mechanistic model of sensory physiology. His research aimed to provide physical explanations for how the brain processes visual and other sensory inputs and integrates that information through the soul. As an artist, he was motivated to study the surface features of human anatomy, showcasing his exceptional skill in draftsmanship.

Could it be that, with certain professions, people would swap healthy limbs for a far more functional artificial one? Would the military find soldiers who could run at high speeds to be an asset on the battlefield? Would electronic eyes with night vision and long-range vision capabilities be the way forward? Would programmable robotic arms give an individual an advantage when working with complex machinery? After all, you only need to download a particular skill set using the Neural Network!

Will our education be delivered much quicker when all we need to do is sit in the download station and receive it in seconds? Will we no longer need huge televisions in our homes when we can have the same experience directly to our bionic eyes?

> *Will we be able to live forever*
> *by just replacing a few plug-in parts?*

There will no longer be a north / south divide for opportunities and investment, nor will there be a separation of the haves and the have-nots. We have the choice of what we want to be and are then kitted out with what we need by our employer. This may include speciality brain chips, implants, or attributes. Whatever you do, don't change your mind. A job will be completely for life!

We will take our holidays and entertainment virtually without leaving the comfort of our homes. We will simply enter our virtual world, go where we fancy, and even go back in time. Both work and leisure will be in virtual spaces.

Getting to this future is already underway. Most of the technology and the corporations that will deliver it exist. All it takes is a choice for us to choose this way of life.

As we see this happening, will you choose to be on or off the grid? Will you choose to follow the growing majority, or will you opt for a far simpler life that you remain in control of?

Take the choices you make today with great care!

THE NEXT ONE HUNDRED YEARS
We can not stop inventing because we must move forward in our evolution, whatever the outcome. Technology, globalisation, human intellectual evolution, communication, and the speed of life are all fast, and there is no end in sight until the last one of us is alive. There are so many things we will see in the future that we do not have today.

Who denies the reason of things, publishes his ignorance.
Leonardo da Vinci

The Empire State Building is being sent to the Moon.

A Special Thanks for Their Contributions goes to:

Aburizal and Tatty Bakrie	As the project patron.
Chris Day	As my publisher and distributor. Certainly the best intellectual contributor through the entire production. A beautiful mind.
Gabriella Brussino	As the writer of those beautiful poems associated to some of my drawings. A constant inspiration for doing always the right thing in all those years we have been together. Page 219, 221 and 223
Pietro Galifi and Stefano Moretti	as my travelling companions throughout the entire evolution of my creative life. I consider myself very lucky to have found these 2 Artists and friends when I was still very young, and I am even more lucky to have them still with me today.
Midjourney Bot	For all those images combining my thinking translated in powerful images by AI.
Teo Mattiozzi	For his contribution to smart thinking, and exchange of concept evaluation and for the cooperation on the creation of Art pieces. Page 57, 58 and 64.
Eliano Mattiozzi Petralia,	For the original music throughout the entire book. As a visionary in the Rossini Mozart connection concept.

Lynn John	For his capacity to question, evaluate, judge and stimulate that type of 'thinking' that goes beyond. Also, for his research and the work we have done together over the years. As the writer of 'Il Maestro' the Rossini and Mozart connection concept.
Pino Confessa	A great partner and friend in the process of looking into our lives. The connection with the past, our existence today and passing it all to the future that will be, but that we will never see.
Shin Ya Yu	Throughout the years, a great visionary to share new concepts with. The Cube to start with. Page 198 to 201
Roberto Tazioli	As the technical team leader, that work together with AVIOAVIO in developing the 'New Bridge' concept. Page 75
Roberto Palmieri	A constant presence and inspiration into 'do it, and do it better'. A Maestro of life.
Gilia Mattiozzi Petralia	As the artistic curator and designer of the 'Leonardo is back' website.
Gigliola Gallo	As my creative partner in the art piece, 'Armonia'. Page 253
Marco Cardelli	As a superb photographer that shoots those portraits placed on the AVIOAVIO Bag design. Page 149
Mariuccia	A very special presence in my life. An elegant, inspiring soul. Page 148

Giamaica Salvato As the AVIOAVIO logo designer
Rodolfo Mattiozzi As a great artist with an amazing natural talent. We worked together in the creation of the Art pieces.
Page 57, 58 and 64.

To know how to listen means to possess, in addition to your own, the brains of others.
Leonardo da Vinci

My enigmatic Mona Lisa

While I thought that I was learning how to live, I have been learning how to die.
Leonardo da Vinci

Published by Filament Publishing Ltd
14, Croydon Road, Beddington, Croydon, Surrey CR0 4PA
+44 (0)20 8688 2598
www.filamentpublishing.com

Beyond Leonardo by Avio Mattiozzi Petralia
© ™PT. Duta Mitra Nawa, © 2025

ISBN 978-1-915465-86-3

The right of Avio Mattiozzi Petralia to be identified as the Author of this work has been asserted by him in accordance with the Designs and Copyrights Act 1988.

All rights reserved.

No portion of this work may be copied in any form without the prior written permission of the Publisher.

Printed in the UK, USA and Indonesia.

A completely new Universe can be created and experienced by everyone.

> Rarely who walk well fall.
> Leonardo da Vinci

> If you are alone, you will be all your own and will own 100% of your time.
> Leonardo da Vinci

The natural desire of good men is knowledge.
Leonardo da Vinci

And just before passing away, Leonardo said: "I have offended God and men because my work has not reached the quality it should have".

Where the spirit does not work with the hands, there is no art.
Leonardo da Vinci

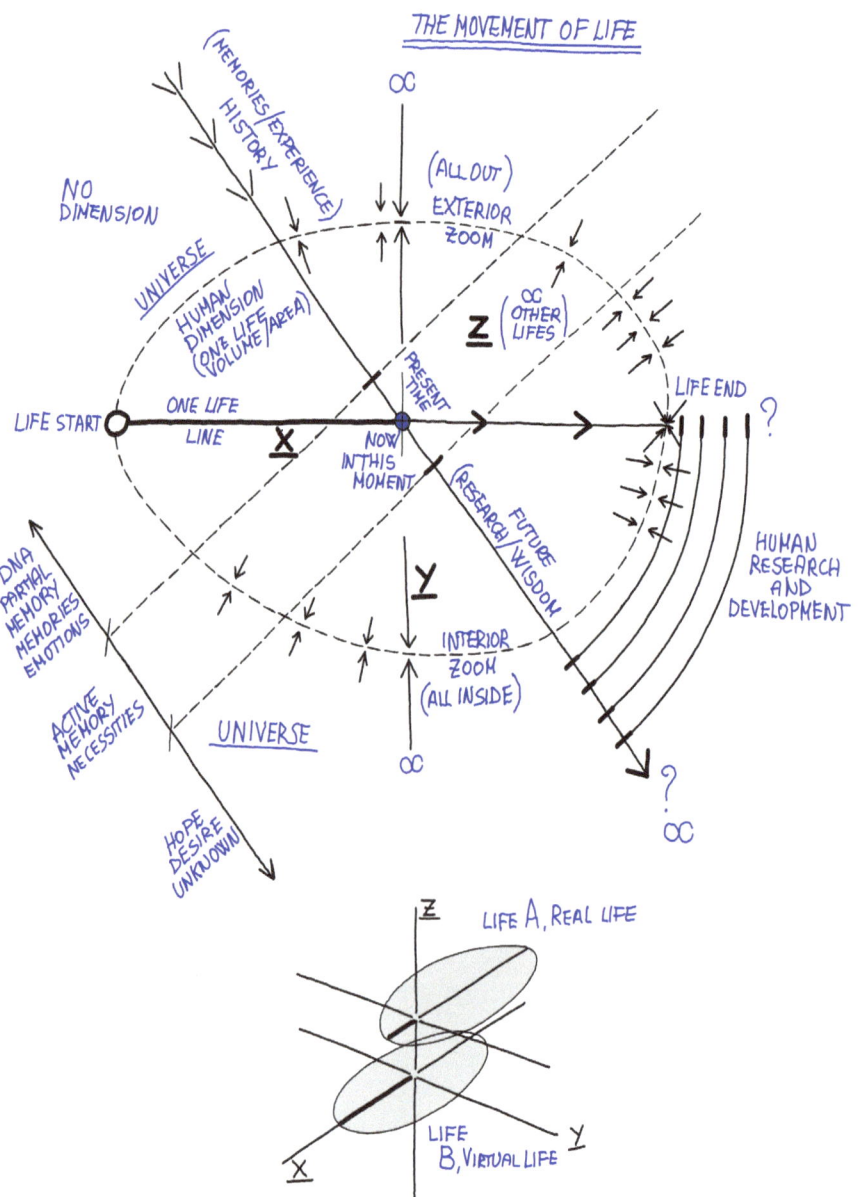

Harmony

I have been impressed with the urgency of doing. Knowing is not enough; we must apply. Being willing is not enough; we must do.
Leonardo da Vinci

www.ingramcontent.com/pod-product-compliance
Lightning Source LLC
Chambersburg PA
CBHW042258280426
43661CB00097BA/1180